How Green Are the Gorons?

Liberal Propaganda Out of Control

Gerald Westbrook

iUniverse, Inc.
Bloomington

iUniverse books may be ordered through booksellers or by contacting:

iUniverse
1663 Liberty Drive
Bloomington, IN 47403
www.iuniverse.com
1-800-Authors (1-800-288-4677)

ISBN: 978-1-4502-7369-5 (sc)
ISBN: 978-1-4502-7370-1 (hc)
ISBN: 978-1-4502-7371-8 (ebook)

Printed in the United States of America

iUniverse rev. date: 01/04/2011

This book is dedicated to my father

Eric Stanley Westbrook

April 14th 1899 — November 7th 1963

My love for nature stems from my father. He was a master gardener. Gladiolas were his specialty, but he also took great care with his Dahlias, Peonies and tomato plants.

He did his best to get his family to the superb parks and lakes of this region, but none of these trips were easy. There were many miles of gravel roads to deal with, always awful, sometimes terrifying. The economic realities of the Depression presented a further challenge for a family of five. There was gasoline rationing to deal with until World War II had run its course. Yet he got us there.

I can't remember the name of the first lake he took me to in the north. There are so many lakes there—Anglin, Christopher, Emma, Sandy— some in the national park and some near the park entrance. However, I do remember the first boat. It was a very small boat, with a very small motor, but it took us to every spot we wanted to go.

I have frequently expressed my awe at the grandeur of the botanical bounty of Earth, whether in the form of flowers, vegetables or grain fields, or on a larger scale in rivers, lakes, parks and forests. All of this came through the teachings from my father. He taught me to love and respect Green, long before Green became a political movement.

Table of Contents

Preface.

• How Green Was My Valley?

This movie[(0)] is one of the greatest movies of all times. This film won five Oscars in 1941 including best picture and best director. It has been called a gentle masterpiece.

What ever is said in this book has no association with that movie other than the wonderful title of that movie to be the inspiration for the title of this book. It in no way is meant to diminish or tarnish the stature of that movie.

• How Green are the Gorons?

This will be a profound question to those familiar with the Gorons and their tyranny in the political, energy, environmental and climate change fields. For those not familiar a brief background is in order. This segment of the space-alien community was first discovered, and reported on, in this writers initial book[(1)]. This book covered, in part, *The Invasion of the Gorons*, an invasion by a band of aliens from the planet *Gore*, intent on increased environmental and climate regulations. The first earthling they captured was Albert Gore Jr. And Mr. Gore proved to be a model student. Indeed his brainwashing was completed in a record time. He has since risen rapidly in the space alien community to become the Head Goron (HG).

There is no question that the Gorons have declared war against the United States, Canada and other countries. But who are the Gorons you might ask? The Goron name surely applies to the ultra liberals, the far lefters and the global warmers. The Goron name is also used as a synonym for:

· those activists/politicians who embrace simplistic science as the answer to complex problems;
· those activists/politicians who utilize complex science to isolate rather than illuminate an issue;
· those environmentalists/warmers who exploit environmental/climate issues for their own purposes;
· those environmentalists/warmers who see *An Inconvenient Truth*, etc, as the gospel; and for

· those activists/politicians/environmentalists/warmers who believe their cause is so righteous that it justifies any means, including deceit, violent demonstrations, sabotage and even terrorism.

• **The Gore Wars**.

Other writers have also commented about this war. A key example is the *Gore Wars* [2] as detailed by Judge Alex Kozinski. In this essay, Judge Alex Kozinski notes the tidal wave of doomsday predictions, started with *The Limits to Growth*. This tome was authored by a group of scientists with the name The Club of Rome. Today, this tidal wave of propaganda continues. Each issue raised is essentially impossible for the public to assess. And they all carry the imprimatur of some scientific-sounding group "ready to vouch that the crisis will cause as much damage to Earth as the Death Star did to planet Alderaan." Politicians jump on board, making the latest crisis the centerpiece of their pitch. This process has led to frequent, costly and in many cases, "disruptive changes in our laws that are difficult or impossible to undo."

• **DeepWater Horizon Incident (DW-H)**.

(1) <u>Prologue</u>. A major theme of this book is a defense of fossil fuels. Not only can we live with a fossil fuel future, but in my assessment, we must live with such a future. Hence, when the DW-H incident occurred it was a deep shock to this writer. Surely such a focus will now be more difficult to implement.

(2) <u>The Latest Crisis</u>. Today, the latest crisis is the explosion on the DW-H offshore drilling platform. This represents a rather outrageous failure of industrial and government responsibility. As one who believes industrial accidents[3] in the hydrocarbon industries should not happen, but they do. We, as a nation, need to examine the evidence very carefully and objectively, which will take considerable time.

However, there are at least three analogies that can be useful on putting the DW-H into perspective.

· <u>Chernoybl meltdown - 1986</u>. The USSR built[4] this plant in "an irresponsibly dangerous way and operated it similarly." This incident led to 60 deaths, and 200 children needing thyroid treatment. Also thousands of unborn children died in Europe via abortion, as their mothers became petrified by the threat from low level radiation (LLR). However, the relatively low risks from LLR were well known by 1986, but you would never have known that via the media. "These deaths were caused by irresponsible media."

· Three Mile Island nuclear incident - 1979. In spite of zero deaths from this incident, the media[4] created a false scenario at this plant, that ultimately led to the end of nuclear power plant construction in the U.S.

· The Enron collapse - 2001 As noted in §6.3 of this book—and Reference 5.2 of that section—Enron's boom and bust had more to do with government regulation than free markets. The fate of Ken Lay, the CEO of Enron, was not the saga of a capitalist wildcatter, but the tragedy of a political rent-seeker.

(3) The Magnitude of this Disaster. The above three examples make the points that the government and the media had major shares in each incident. Yet for the DW-H all we seem to hear is BP, BP, and more BP. And the environmental movement[5] can and will use this incident as a major rallying cry. Indeed this reference noted the infamous quote by Rahm Emanuel: "a disaster is a terrible thing to waste." And most of the media[6] is engaged in an effort to breathlessly tell the viewers and listeners that this accident is causing an ecological disaster of epic proportions." It is possible that verdict is a bit premature. Rush Limbaugh, in a transcript[7] of his June 1, 2010 program, presents a chart[8] that compares the DW-H release versus six other sources. The comparison is stark. While the DW-H is shown larger than the Exxon Valdez, but smaller than all others and much smaller in some cases. Surely the DW-H incident[9] is bad, "but experts say magnitude unclear." These writers also noted: The White House is ignoring all the shades and complexities here to make a dramatic point. They also noted: "the impossibility of ranking such a varied list of catastrophes." And finally they observe: "Perhaps the worst disaster is always the one people are living through now."

(4) Epilogue. Then we have politics. Not only do the Gulf residents and businesses have the oil leakage to contend with, but also the Washington politicians and friends. Some examples follow.

· A six month deepwater moratorium. When this was announced[10] , Obama "pointed to an Interior Department report of new 'safety' recommendations." He reported that these recommendations, including the six month drilling ban, had been reviewed by a wide range of other experts - - - "the nations drilling brain trust." The opposite was true. Eight of the experts the administration had listed in its report said their names had been used - - - .And the draft report they reviewed "had not included a six month drilling moratorium."

· The President's Animosities. This editorial[11] highlights the level of animosity that prevails between the President and industry. The editorialist reported: "I cant recall any previous president with this depth of visceral, anti-business animosity."

· BP's Truce Could be Short Lived. With the agreement[12] to set up a $20 billion escrow fund, "the optimistic reading is that BP has negotiated a truce with the White House at a manageable cost." Don't bet on it!

· The BP Precedent - Who's the bigger scoundrel? The people seem to have concluded[13] "that they dislike both BP for causing this disaster, and the White House for appearing to exploit it." - - - "It's hard to know who is more unlovable, BP or its Washington expropriators."

There have been several inputs that the Obama administration are holding the whole gulf coast region hostage as a means to pursue their energy-climate change objectives. It is clear that this political and legal struggle will go on for a very long time. In any event, this will not change the focus of this book.

• *Vision of the Anointed*.

The techniques used on the energy, environmental and climate change issues are not unique to that field. They are applied to all of the major political issues of the 20th century. Thomas Sowell noted [14], in his *Vision of the Anointed*: "What all these highly disparate crusades have in common is their moral exaltation of the anointed above others, who are to have their very different views nullified and superseded by the views of the anointed, imposed by the power of the government."

Sowell reports that several key elements are common in all such movements:

· "Assertions of a great danger to the whole society, a danger to which the masses are oblivious."

· "An urgent need for action to avert impending catastrophe."

· "A need for government to curtail the behavior of many, in response to the conclusions of the few."

· "A dismissal of arguments as either uninformed, irresponsible or motivated by unworthy purposes."

Yet with each subsequent, trumped up crisis, the press provides usually unlimited, un-researched, and un-skeptical support. Serving as a shill for these movements is not journalism. It is a corruption of the profession. It is prostitution, a selling out of the values of honest journalism for the adoration of the so-called *elite*, and the hope of ultimately becoming one of these *so-called elite*.

• *Useful Idiots.*

While my book[1] implied there may be *Morons* involved here, Mona Charon's book was a bit more direct, as it was entitled the *Useful Idiots*. The *Idiots* endorsed anything and everything that the communists said or did [15]. On the book cover she flagged "Jane Fonda, Dan Rather, Ted Kennedy, Jimmy Carter, Jesse Jackson and all the other liberals who were and are always willing to blame America first, and defend it's enemies as simply 'misunderstood'."

She argued that these are the same liberals who flocked to Cuba and called it paradise, just as earlier liberals visited the USSR and proclaimed its glorious future. Other liberals cited included Madeline Albright and Strobe Talbert all of whom turned a blind eye to the Soviet 'Evil Empire'."

More recently, Glenn Beck picked up on this theme in his colorful book[16]: *Arguing With Idiots.*

Then in the morning edition of the NY Times for 9/11/01—in a most astonishing example of timing—this propaganda machine "published [17] a sympathetic portrayal of two American terrorists who had been responsible for several bombings in the 1970s." These were members of the Weather Underground—a radical offshoot of the already radical new left. The terrorists were Bill Ayres and his wife, Bernadine Dohrn. While their bombs:

· 1970: NY City Police headquarters;
· 1971: U. S. Capital building; and
· 1972: the Pentagon did not kill any one, they were outrageous and pathetic, and represented a harbinger of events to come. How Ayres and Dohrn came to the conclusion that such bombing was a positive thing to do is rather mind-numbing. How the Times— this nations so-called *newspaper of record*—saw fit to publish Ayres opinions such as he finds "a certain eloquence to bombs, a poetry and a pattern from a safe distance" is even more mind-numbing. And how Barack Husein Obama (BHO), came to the judgement[18] that Ayres was someone worthy of visiting and associating with. More on this later.

Finally on the back cover of Charon's book, William Kristol, editor of *The Weekly Standard* commented: "With America once again at war, its useful to be reminded of the phenomenon of *useful idiots*. "Unfortunately they are once again with us."

• Conclusion.

If one equates the *Gorons* with the *Idiots,* then the *Gorons* presence in our society is even stronger than previously depicted. Hence the questions on their greenness becomes important. Are the Gorons truly green, or are many of them simply cloaking themselves with a green robe as a ruse to gain and hold office? This book will address that question.

About the Author.

Mr. Westbrook and his wife were both born in western Canada: Saskatoon and Calgary respectively. He and his wife moved to the United States in 1958. They are naturalized American citizens. Their three children were born in the USA.

The writers credentials includes a B.Sc. degree in chemical engineering from the University of Saskatchewan, with great distinction; an M.Sc. Degree in chemical engineering and an M.A. degree in energy economics from the University of Minnesota, each with a minor in mathematics.

The author has worked extensively, form 1955 to 1994, in the hydrocarbon industries, first with an oil company and then, five years later, with a petrochemical company. He has obtained additional experience in this field with a wide variety of consulting assignments. Overall this experience has focused on a very broad set of energy problems and issues, including alternative energies. It has also included many water supply, water quality, waste water management, other environmental problems and the whole spectrum of climate related problems and issues.

On retirement Mr. Westbrook has remained active in consulting, educational services, public affairs and a retailing activity. The author has served for several years as a senior associate, initially at the University of Houston and now at the Center for Energy Economics, at the University of Texas in Houston. His background hence includes an industrial, educational and free enterprise perspective.

The author also has 70 plus years as an observer to nature, agriculture, research, civil service, industry, politics, and society at work in five distinct geographical and geopolitical arenas of North America: Saskatchewan, Ontario, Minnesota, Michigan and Texas. He has published dozens of papers, essays, editorials and letters—on energy, environmental, climate and political issues—and one book.

He rejects the position that since he worked in the above industries, he cannot protest what is going on in the environmental, regulatory and political fields. He will be charged with having a vested interest. Financially this is trivial, with zero shares in the petrochemical firm, and 100 shares in the oil company.

Philosophically Mr. Westbrook admits he would be more inclined to hear out industrial companies than environmental activists would be. However, he notes that he frequently operated as a maverick at these companies, and his relationships with both companies was not always in sync. Finally he would argue the average environmentalist—who has staked out his, or her pathway in life on the environmental movement—has a far bigger vested interest in getting their way on environmental, climate and energy issues.

A few more words are appropriate for the issue of not always being in sync with his employers. At the petrochemical company, part of this was due to starting his career there essentially as an internal consultant, a role bound to create some friction. Part of this was also due to literally having no *functional home* during this period. While formally located in manufacturing, over half of the work he did was on loan to the marketing and business areas.

After seven years in that role he perhaps should have moved on to an external management consulting firm. He had also been offered a visiting professorship for the coming school year, by the University of Texas. However, due to an attractive offer to join the major business area of the petrochemical firm, he turned the University of Texas down and stayed, for an additional 27 years, at the petrochemical company.

Over this time he often found it difficult to forget that he was no longer acting in a consulting role. Many times he would find myself in situations where he had strong differences of opinion with the actions planned or undertaken. He was no yes man. Indeed, his behavior at times, was that of a maverick.

Acknowledgements to two key professors.

These professors both obtained their PhDs from
the Kings College, University of London.

• **J.W.T.Spinks** (JWT).

JWT became an assistant professor of chemistry at the University of Saskatchewan in 1930. During 1933 - 1934 he headed to Germany to study under Gerhard Herzberg, a specialist in molecular and atomic spectroscopy. Spinks returned to Saskatchewan in the fall of 1934, and Herzberg joined him a year later. He served this university for ten years as a professor of physics. In 1971 he won a Nobel Prize.

JWT was head of the Department of Chemistry and Chemical Engineering, and Dean of Graduate Studies at the University of Saskatchewan, when I started there in 1951. He later rose to become president of this university, a position he held for 16 years.

JWT's specialties included radiation chemistry. He was one of the founders of this field, and pioneered the use of radioactive tracers in agriculture. In 1982 he was inducted into The Saskatchewan Agricultural Hall of Fame in part for his work on the improvement of fertilizers.

During WWII he was assigned to the Canadian Operations Research effort, an activity that used logic, mathematics and statistics on operational problems. Later JWT became part of the Canadian atomic energy project. His interest in, and confidence on nuclear power contributed towards my interest in this energy alternative. However, his activities in Operations Research also created a major interest for me in that field, an area that I focused on at graduate school and later.

• **Rutherford Aris**.

I attended Minnesota from September 1958 to June 1960 and obtained an M.S. in Chemical Engineering and an M.A. in Energy Economics, with Aris as my advisor on each thesis. I look back, with a great deal of fondness, to studying under his leadership.

The two thesis that I was involved in at Minnesota were on an analysis of chemical reactor safety/stability, and on energy conservation in the process industries.

References and Notes

(0) Movie *How Green Was My Valley (1941)*.

(1) Westbrook, Gerald T., *'Acid Rains' on Liberal Propaganda, Ultra Liberals, Far Lefters and Global Warmers Beware*, iUniverse, Lincoln NE, December 9, 2004. In the main text this citation referred to the *space — alien* community. This consists of *Gorons, Gluons, Muons and Morons*. The *Morons* originated on planet *More*, the neighboring planet to planet *Gore*.

Since *Gorons* and *Morons* are so much alike it is essentially impossible to tell them apart. In this paper the *Goron* name will be primarily used, but the reader should be aware that anytime the *Goron* name is used, it is very possible that, that *Goron*, might in reality be a *Moron*. Finally it should be confessed that Gluons and *Muons* are actually nuclear particles,

(2) Kozinski, Alex, *Gore Wars*, Michigan Law Review, August, 2002, Pages 1742 - 1767. Judge Alex Kozinski sits on the U. S. Court of Appeals for the Ninth Circuit.

(3) My experience with safety in industry in general, and with explosions in particular, include:

· a tiny contribution to a research effort on grain dust explosions at a national laboratory in Canada;

· a masters thesis on chemical reactor stability; and

· participation in hundreds of safety meetings—at the petrochemical company I worked at for 34 years— where safety was a dominant priority, and an employment termination possibility.

(4) Robinson, Arthur B., *Accidents*, Access To Energy, April, 2010.

(5) Galbraith, Kate, *Environmentalists Use Oil Spill as a Rallying Cry*, New York Times, June 13, 2010.

(6) Robinson, Arthur B., *Oil Drilling Accident*, Access To Energy, May, 2010.

(7) Limbaugh, Rush, *This Oil Disaster in Perspective and a Reminder of Saddam Wells*, Transcript, The Rush Limbaugh Show, June 1, 2010.

(8) The chart in the above reference, that compares seven oil spills, was provided by Roy Spencer, University of Alabama at Huntsville.

(9) Gillis, Justin, It's Bad, but is spill the worst disaster?, New York Times, as reported in the Houston Chronicle, June 19, 2010

(10) Editorial, *Crude Politics*, The Wall Street Journal, June 17, 2010.

(11) Henninger, Daniel, *The President's Animosities*, The Wall Street Journal, June 17, 2010.

(12) Denning, Liam, *BP's Truce Could be Short-Lived*, The Wall Street Journal, June 18, 2010.

(13) Editorial, *The BP Precedent - Who's the Biggest Scoundel*, The Wall Street Journal, June 18, 2010.

(14) Sowell, Thomas, *The Vision of the Anointed*, Basic Books, A Division of Harper Collins, Inc., New York, NY, 1995.

(15) Charon, Mona, *Useful Idiots - How liberals Got it Wrong in the Cold War and Still Blame America First*, Regnery Publishing, Inc., Washington DC, 2003.

(16) Beck, Glenn, *Arguing With Idiots: How to Stop Small Minds and Big Government*, Threshold Editions - Mercury Radio Arts, a Division of Simon & Schuster, Inc., New York, NY, 10020, 2009. (17) Smith, Dinitia, *No Regrets for a Love of Explosives; In a Memoir of Sorts, a War Protester Talks of Life With the Weathermen*, New York Times, September 11, 2001.

(18) Moran, Rick, *Obama and the Radicals*. This has been posted on "the Two Malcontents" site.

See also : http://technorati.com/blogs and www.the_two_malcontents.com/page/9/, February 27, 2008.

1. Introduction.

1.1 Background.

In my first book I warned the world about the *infiltration of Gorons into our society*. "These aliens have gained control of most of the media, most of the educational systems and most of the entertainment outlets." I warned the world about the *Head Goron (HG)* and his crusade. And I warned the world on his extremism[1.1]. He is drawn, not to the most reasoned scenario, but to the most extreme. I noted then that, while the political future of the HG looked pretty bleak, the Goron threat would live on.

Now Albert Gore Jr. has received enormous publicity for his movie and new book[1.2] *An Inconvenient Truth*. Their success surely amplifies the threat from the Gorons. Indeed with their victory in the 2006 and 2008 elections, and the 2007 Supreme Court vote, the Goron threat will be far stronger

In my humble opinion, the world needs an antidote.
· If you would like inputs on some of the extremist thinking and writings by the *HG*;
· if you need relief for all the distortions and hype he has used on the global warming (GW) issue;
· if you would like a perfect squelch for the views of this politician and others of his ilk;
permit me to offer my first book, and indeed this second book, for such roles.

One of the key features of 'Acid Rains' (AR) is the use of satire to help describe, define and deflate the Gorons. Some of this satire is pretty acidic, hence the title of that book. It is acid satire that rains on liberal propaganda.

'AR' was triggered by the writings[1.1] of Albert Gore Jr., which are shown in to be nothing more than propaganda. This sequel to 'AR' will amplify on the propaganda efforts by the HG and others, such as Speaker of the House, Nancy Pelosi (NP). Clearly we still live in a deep ocean of propaganda. We will need many searchlights focused on this situation—to overcome this propaganda fog and bring visibility to the public—if we are ever to get it righted. As with the original book, this sequel will attempt to contribute to that objective.

1.2 The Pelosi Syndrome.

• Alliteration around the Letter P.

I will be referring to many words and names in this section that start with the letter P, including Ms. Pelosi. Perhaps it was Frank Sinatra, in his song *That's Life*, who was the first to highlight the alliteration on the letter P. He sang of "a puppet, a pirate, a poet, a pauper, a pawn and a king." He also sang about a *Nancy With The Laughing Face*, but that couldn't have been about NP.

Yes, I too have talked and written of similar alliteration on this letter P, but surely not as glamorous as Sinatra. My writing was based on my concern on propaganda. I have noted in my initial book that there is no question that the Gorons have declared war against the United States. Most wars come complete with propaganda campaigns, and this war is surely no exception. In the course of assessing such propaganda, I have "sang" of **pimps and prostitutes and propagandists.**

Propaganda comes in all sizes, colors and shapes, spread by the pimps and the prostitutes involved. Part of such campaigns includes news manipulation and slanted messages. Manipulation in the print media includes selection, or non-selection, of specific news items and their relative treatment:
· headlines and sub-headlines;
· space allotted;
· editing - removing sentences, even segments;
· editing - adding sentences, and comments and
· placement - inputs in support of the current issue of the day gets front page treatment, any corrections will · placement - any corrections will be buried deep in the paper.

• Are there Pimps and Prostitutes and Propagandists at Work at Newsweek?

There are so many pimps and prostitutes involved in these campaigns, striving to attract the uneducated, the naive and the uninitiated to their political agenda. One of the more egregious of all propaganda blasts came in Newsweek[2.1] where the cover blared "Global Warming is A Hoax*" - This was as a vile, venomous piece of writing masquerading as journalism[2.2]. Note, the asterisk here refers to a footnote on the cover which states: "Or so claim well-funded nay-sayers who still reject the overwhelming evidence of climate change." This latest writing entitled *The Truth About Denial*, by Sharon Begley et al, has, as openers a cover displaying a fantastic and terrifying picture of the sun. As such, one might think these writers were leading supporters of the alternative theory on global warming, namely natural climate variation due to solar and astrophysical phenomena. Not hardly. They are hard-core greenhouse gas advocates and society-is-guilty cheerleaders.

Begley is/has been the science writer for *Newsweek* and the *Wall Street Journal*. She has authored high powered books on science and health. As such, one might think she would have respect for the scientific method: observe; hypothesize; predict; test predictions, then modify hypothesis as seems appropriate. Repeat process until satisfied, then promote the hypothesis to become a theory. The great advantage of this approach is that one does not have to believe a given researcher, one can redo the experiment and establish if the researcher's work is valid.

One would also think Begley would have respect for the creed by Thomas Huxley (1825 - 1895): skepticism is a scientist's greatest responsibility; blind faith is his greatest sin. He was a passionate defender of Darwin's Theory. However, he may be best known for his famous debate, in June 1860, with Archbishop Samuel Wilberforce. During this debate, Wilberforce ridiculed evolution and asked Huxley whether he was descended from an ape on his grandmother's side or his grandfather's. Huxley's response: "If then, said I, the question is put to me would I rather have a miserable ape for a grandfather or a man highly endowed by nature and possessed of great means of influence and yet who employs these faculties and that influence for the mere purpose of introducing ridicule in to a grave scientific discussion, I unhesitatingly affirm my preference for the ape."

In any event this Newsweek writing triggered my commentary[2.3] on *More on Propaganda as Journalism, including the "P, P & P Test"*, a test for pimps, prostitutes and/or propagandists. Surely there is a major part of the Pelosi Syndrome at work here.

• The Entertainers

In addition to editors and journalists, perhaps the next most important group are entertainers: actors, singers and comics, with singers being the most vocal of the lot. These are the ones who have, for example, fantastic vocal talents, but who have used the fame gained from such talents as a pathway into the minds of the uneducated, the naive and the uninitiated. I had mentioned Barbra Streisand (BS) in my first book. In her latest effort, good old BS has become an expert on hurricanes. As I write this chapter a new Goron has emerged, one Sheryl Crow. Clearly she is an expert on global warming. While she attempted to engage Karl Rove in discussion of her views on this subject, this effort was not successful. Which is a good result, as her key recommendation appears to have been that individuals should limit themselves to one square of toilet paper per visit. I am sure we can all agree that that form of adaptation to the GW crisis is one that we can all readily salute and embrace. Profound? Well not hardly.

• The Media for Women

Today all of us are besieged daily with thousands of messages, not much more profound than that offered by Ms. Crow. These messages come from TV, radio, newspapers, magazines, the Internet and so on. Magazines are endemic. Not only do they come in the mail or in the magazine sections of book stores, but certain types of magazines are stuck in the face of every housewife and teenager at every check out station in the country. The magazines offer[3] secrets on "How to fight stress", secrets on how to stay thin, and the "Secret Sex Move No Man Can Resist." "And on and on and on.".

This former editor of the *Ladies Home Journal*—a relatively tame magazine by check-out island standards—went on: "I know from long experience that media for women tells you endlessly about stress in your life, about the way you should look, about what should make you feel sorry for yourself, or very, very fearful about your health and the environment."

The business of women's magazines and TV is primarily based on telling women that they are being victimized. It tells them their lives "are often too tough for them to handle and that they should feel very sorry for themselves." This distorted vision is rather incredible when one recognizes one is part of "the best educated, healthiest, wealthiest, longest lived women with more opportunities for personnel fulfillment than any other generation in history."

This former editor went on: You are repeatedly given the "favorite celebrities liberal messages, with a halo of approval." In much the same way "you are given a one-sided message about politics, by always being told more government is the best solution to fix any of the problems of your life." More government, more government and more government. That solution applies to every area of life, including education, medical, transportation, the environment and the global warming issue.

In such a fashion the media for women seeks to con or snow their audience. Would you believe brainwash. But this objective applies to all media. Without the understanding that we live in a daily propaganda tsunami, individuals will become easy marks for the industrial, commercial, environmental, educational and political shysters or pimps that are now endemic in our society.

• The Heather Mallick Affair.

I thought the Newsweek attack was the worst one would ever see. How wrong can one be, for on September 5th, 2008, a Heather Mallick, blessed the world with her commentary[(4)] on the Republicans VP selection. Some highlights of this mind numbing hatchet job follow. The low-lights will not be quoted.

· "Palin was not a sure choice, not even for the stolidly Republican ladies branch of Citizens for a a Tackier America. "No, she isn't even female really."

· She asks why Sarah Palin?, as the Republicans have already sewn up the *white trash* vote. She gives us her definition here. "White trash—not trailer trash, that's something different—is rural, loud, proudly unlettered (like Bush himself) suspicious of the urban, disbelieving of the foreign, and a fan of the American cliché of authenticity."

· Palin has toned-down version of the porn actress look - - - ."

· Palin has it all, along with being vicious and profoundly dishonest."

· The Palins "claim to be family obsessed while being tudiously terrible at parenting."

· Finally Mallick claimed to have "an attachment to children that verges on he irrational, but why don't the Palins?"

Now it is easy to understand attacks on Sarah Palin from the American left. They seem to be petrified that she might head up the GOP ticket in 2012. Absolutely petrified. So they are doing their utmost to discredit her now, humiliate her now and eliminate her now. Bur Mallick has no vote, so why such an attack? It took over three weeks, but CBC News finally apologized for this column, conceding it was "a classic piece of political invective. "It is viciously personal, grossly hyperbolic and intensely partisan." It also conceded "that many of her most savage assertions lack a basis in fact." Yet it was still published.

This ugly diatribe motivated this writer to seek broader circulation of the P, P & P Test. A version was sent to the editor of the Canada Free Press, a Ms. Judi McLeod, but it was not published.

Next I tried another version on the Sean Hannity Discussion site, which led to the following message: "You have been banned for the following reason." However, no reason was specified. Their next line informed me on "the date the ban will be lifted." And their answer: "Never." Well, I must admit their response got my attention. My oldest daughter has repeatedly counseled me to strive to emulate Charles Krauthammer and not Dennis Miller. Perhaps she has a point.

Well I was about to give in when I noted an apology by Greta Van Susteran. She had called Mallick a pig and she deemed it appropriate to apologize. However, this apology was not to Mallick, but to all pigs. As a result I sent off another version of the P, P &P commentary and it was posted.

• The Chief Male Propagandist - the HG

But who is the chief male propagandist you might ask? Would you believe none other than the HG himself. The HG served as a journalist during the Vietnam War. Not only did the HG develop an interest in journalism during this war, but also in propaganda, as is apparent as one reads his first book. Today, he has brought the practice of propaganda to new heights. In a commentary on the HG's efforts at Columbia University, following the 2000 election, Joseph Farah reported that "Al Gore is not teaching journalism. He is teaching propaganda[5]." This was in reference to the instructions given to his students on how to prepare a critique "on media coverage of the GW issue."

His new book and movie are superb efforts at brain washing. The HG has progressed from a simple, black and white propagandist in his first book to one of high technicolor. His new book uses, or more precisely abuses color to a high degree, particular the color red. This abuse of color on hurricanes and ice-sheets is covered in some detail in Chapter 2.

In any event the political career of the HG has been revived. Many see this as a reincarnation. Some see it as *a new coronation of Prince Albert.*

• The Chief Female Propagandist - the NP.

(1) <u>Her Coronation.</u> But who is the chief female propagandist you might ask? I would nominate none other than the NP. A simple search—for "Nancy Pelosi" and propaganda—yields almost 400,000 citations. Now many of these may well be from supporters, but it is a start at making the claim that she is the chief female propagandist. One of these citations[6]

reported that NP's "elevation to the speakers chair just two years ago had its own elements of a coronation - - - ."

As part of this coronation her supporters came up with commemorative buttons for her swearing-in, "depicting her as Rosie the Riveter flexing her muscles." That just might be the biggest portion of bs ever served up to the American public. And to the extent that bs is propagandistic, this button is propagandistic. Now it is possible she had absolutely nothing to do with these buttons, but not hardly. Whatever issues Mrs. Pelosi may claim her leadership has defined the direction. "Her intransigence has set the tone. "And her penchant for excess - ." In short she is involved everywhere.

Now as part of her elevation to the Speakers office, NP stated [7] that "100 hours time enough to begin to 'drain the swamp' after more than a decade of Republican rule." I too have written[8] of the incredible biological activity in DC, albeit on the Senate side. I made the point: that primordial ooze had been created on the left side of the Senate.

I rather doubt we will be able to resolve which party and which side of congress is more responsible for this swamp in DC. We all should be delighted with Ms. Pelosi's objective. However, the idea that the largest swamp in all of history could be drained in 100 hours, has got to be the ultimate in hype, if not in propaganda. Early in 2009 an editorial[9] asked "Is the House swamp Drained Yet?" This editorial, while reporting broad concern, particularly flagged democrat Charles Rangel.

(2) The Stimulus Package. This "Democrat has become the mother of all stimulus packages." As such she undoubtedly was very pleaseed when the house approved a $818B package, in late January, 2009, albeit this issue rapidly became the property of the President. The final vote was 214 to 188. All Republicans present voted against this bill, as did 11 Democrats. The Senate, in turn, passed an $838B version in early February. And our president signed a $787B final bill on February 17th, 2009.

Charles Krauthammer calls[10] Obama's signature bill "not just bad, not just flawed, but a legislative abomination." Much of the blame for this has to fall on the NP as Obama "inexplicably delegated the writing to Nancy Pelosi and the barons of the house." And NP, who bragged about 100 hour swamp drainings, was surely responsible for much of "the greatest frenzy of old-politics influence peddling ever seen in Washington."

(3) Global Warming. In April 2008, Al Gore and his climate change awareness organization, *The Alliance for Climate Protection*, launched [11] a $300 million, three-year campaign to teach "people in the US and around the world that the climate crisis is both urgent and solvable." This "We"

campaign " utilizes celebrities including Nancy Pelosi to help spread the propaganda that the AGW issue is both urgent and solvable. So NP and the HG have a mutual commitment to the AGW issue.

They have other things in common too, such as both being members of royalty, and both being multi millionaires, albeit the NP's 2007 disclosure reported her families net worth as between plus $86 million and minus $9 million, perhaps a new level for the accounting practice.

She has claimed [12] that the "technological advances that will achieve energy independence also will help us address the most urgent environmental issue facing us today: global warming." In Elsewhere I have written on energy and noted that achieving "energy independence" is an impossibility. Indeed it is not even the correct objective. Hence, to this writer, talk of "energy independence" contributes to the propaganda that we don't need fossil fuels.

While surely not as verbose as the HG—at least on the global warming issue—the NP has still had much to say on this subject. Even in 2007 she declared that "Climate change is already having profound effects on human and biological systems; to avoid catastrophic climate impacts we must start cutting global warming pollution immediately." She even went on to parrot[1,2] the propaganda via the HG re "the loss of major ice sheets and sea level rise affecting hundreds of millions of people." In this book see Sections 2 on the HG and Section 4 on sea level rise.

Finally the NP—wrapping up just about every issue possible into one modest statement—claimed "that the only way to save the economy is to stop climate change, which will cost us untold trillions in mitigation, adaptation, health and security costs if it continues unchecked. Perhaps it is time to call out those opposing clean energy and emissions reductions (and lying about the costs, too) as anti-growth, weak on security and actively working against maintaining public health."

The house bill was passed[13] at the end of June, 2009 "by a 219 to 212 vote, with 44 Democrats defecting and only eight Republicans joining the majority."

(4) <u>Health Care</u>. George Will, perhaps the countries greatest baseball analyst, hit a home run when he summarized[14] the arguments on health care. In one breath Barak Obama says "Medicare is an exemplary program that validates government's prowess at running health systems." In a second breath he also says "Medicare is unsustainable and going broke, and that he will pay for much of his reforms by eliminating the hundreds

of billions of dollars of waste and fraud in this paragon of a program, and in Medicaid."

NP got the stimulus bill through with a 26 vote victory and the cap & trade nightmare by a seven vote margin. Health care—which includes such beauties as expanded coverage, medicare cuts, cost controls, end-of-life issues, single option, funding of abortions, funding of contraception and eugenics—may not get through at all. She has stated[15] that "health care reform would pass the house and that it would include a public plan" these maybe simply a part of her propaganda mantra. Some still think a plan will get through, but it likely will be unrecognizable from NP's initial efforts.

Now NP was not the only operative to push the health care initiative aggressively. BHO bet all his political capital on this issue. Krauthammer[17] reported "the final act was carefully choreographed and included many misrepresentations. It was claimed that this bill contained tort reform. It did not. It also included terrible misrepresentation on it's ultimate cost. This drove Warren Buffet, a one time Obama supporter, to declare "the current senate bill - - - is a failure because the country desperately needs to bend the cost curve down and the bill doesn't do it."

Rumors[18] about the House passing this legislation using the so-called *Slaughter Solution* (*SS*) were wide spread before the vote. NP's "admission that the *SS* is one of about four different options the House is considering in order to pass healthcare reform", indicates it is time to understand the *SS*. This would work as follows:
· BHO would sign the Senate version of the bill;
· the House would pass *fixes* to that bill and would *deem* the Senate bill as passed;
· the Senate would then have to *reconcile the House fixes* and incorporate them into the final bill.

This would allow any Democrat to claim that they never voted for the Senate bill, hence providing them cover. The NP sees this as "an attractive option because it will protect House members politically who are unwilling to support the bill publically."

NP knows "that this bill is unpopular, she knows that the American people are opposed to this bill, yet she is willing to stoop to this level to pass it." And "she thinks the American people are so stupid that once the bill passes and is law that we will not realize that by not voting directly for or against this bill they actually voted for it." The writer concluded that he could not "begin to describe how obnoxious and arrogant" NP's position was on this issue. Perhaps these traits were best revealed[19] in her unbelievable statement, where she claimed "we must pass the health care bill so that we can find out what's in it."

In any event, the Senate health legislation was passed late on Sunday, November 21, by a 219 to 212 count, a 7 vote margin. However, unlike the global warming bill, that was passed in June 2009 by the same margin, with support from eight Republicans. This Health bill had zero republican support. In short it was a 100 percent partisan bill. It included [20] 2409 pages for the Senate passed bill, and 153 for the reconciliation bill.

The reader is referred to the Wall Street Journal (WSJ) reference[20] for much of the detail in this bill. A second WSJ article[21] provides a brief listing of the history for health care legislation, ranging from 1934 up to the present. In a third WSJ commentary[22]. the writer described this bill as deeply flawed. Even though NP backed off the "deem and pass" idea, she still called the process as one of sordid politics. "The final days were a simple death watch, to see how the votes would be bought, bribed or bullied, and how many congressional rules gamed, to get the win." Perhaps Michigan Rep. Bart Stupak was the most egregious prostitute of the lot, as he "sold his anti-abortion soul for a toothless executive order." Finally an editorial [23] noted this is a landmark of liberal governance whose price will be very steep." And the "Democrats have taken responsibility for what comes next."

• Conclusions on the Pelosi Syndrome.

There will be a multiplicity of legal challenges to this healthcare law. These will range all the way from lawsuits up to a call for a national Constitutional Convention. Surely it is far too early to see what, if anything, comes from these initiatives.

As for NP, this victory will surely give her visions of grandeur. However, one cannot be sure how NP sees herself. It is very possible the royalty fixation will set in. Yet the analogy to, say Queen Victoria or Queen Elizabeth, does not seem to be a particularly good fit. The movie queen analogy may be better. Indeed I see her now as a very good analogy to both the star of *Sunset Boulevard* [24], namely Gloria Swanson, and to Norma Desmond, the legendary silent film diva, and the subject of this movie.

I have strived to make the case here that Ms. NP is the chief female propagandist. Now some may argue that what she does is more hype than propaganda, but I think not. Whether it is: hype or hypocrisy or duplicity or hubris or propaganda it is Ms. NP. One might call it The Pelosi

Syndrome – namely one who deals in hype, hypocrisy, duplicity, hubris and propaganda.

1.3 Global warming, Environmental Issues and Political Battles.

It is my conviction that the political debates in general, and the debates over environmental and climate change issues in particular, have been essentially *kidnaped* by the Gorons. This has resulted in a covert shift in emphasis and direction towards political activity and political control as the objective. Rather than a systematic effort to prioritize and eliminate environmental problems we see efforts aimed at political domination. Rather than the development of a comprehensive climate theory, we get endless fear-mongering. The Gorons are not out for science. They have no concern about preserving the scientific method. Put very bluntly, they are out for power.

As noted above, invasions inevitably include propaganda. This writer would argue that our society is under the most <u>intense</u>, <u>crafty</u>, <u>creative</u>, and <u>cunning</u> propaganda campaign ever conducted in the western world. In contrast, the campaign launched by the Nazis, over World War II, literally pales in comparison. The principle propaganda thrusts frequently originate at dozens of different Non Governmental Organizations (NGOs). The reader will be more familiar with such names as the Environmental Defense [Fund] and Greenpeace and so on. And each of these NGOs, press their attack forward against industry, science, energy, chemicals, world trade and free enterprise. In addition many politicians, government bureaucrats, university professors and entertainment personalities all chip in, but all primarily supported via inputs from one or more NGO.

While decentralized, one gets the feeling this propaganda effort is well orchestrated, by an invisible, but all powerful central command — the Propaganda Ministry or the Goron Central Command. Today that conviction is even stronger.

I had noted in my first book that Senator Kerry's wife had organized a foundation, The Tides Center of Pennsylvania, that had funded many very far left wing environmental projects. Mrs. Kerry also had a relationship with the original Tides Foundation and Tides Center of San Francisco. It was noted that these groups appeared to be, among other things, a charity laundering scheme.

I had also noted in my first book that Senator John F. Kerry had "some rather strange people he has supported or embraced." However Kerry was a rank amateur in this field compared to BHO and friends. Indeed some of the NGOs close to the president have metamorphosed into a plethora of almost unrecognizable shapes, intersections and interactions. The Acorn group and the SEIU (The Service Employees International Union) come to mind here.

Finally we have the Czars. Czars to the left of me, Czars to the right of me, here a Czar, there a Czar, every where a Czar, Czar. In most cases their role is undefined or ill defined. Ditto for their power. Are their appointments constitutional? What about the vetting process? Who knows? Surely it has not been used across the board. And one writer asks[16] the question: "was the White House simply too incompetent to properly research [Van] Jones in the vetting process?"

Thus this war with the Gorons and the Czars, goes on and on, with no end truly in sight. As noted in my first book we must be engaged at every new offensive, if we are to avoid the Gulag. It is hard to amplify on that position, but somehow we must, or we will surely be headed to the Gulag.

Perhaps the words[17] of David Horowitz—the son of two life-long members of the Communist Party, but became a staunch conservative writer early in his life—capture our predicament: "Our country is under ssault by a determined, deceitful and powerful left which will stop at nothing to realize its goals. Facing them, I would rather have Glenn Beck out there fighting for our side than 10,000 David Frums who think that appeasing leftists will make them think well of us. No it won't. It will only whet their appetite for our heads.

Now Frum—a former special assistant to President George W. Bush—has fine academic and journalistic accomplishments. He has also received some impressive accolades. Hence he should not be treated so harshly and bluntly. However, I believe what Horowitz is saying here is that today the situation we are in demands a *hound-dog* like Beck to track down and *sniff out* the facts. Further, the situation we are in also demands a *"Junkyard Dog"* like Beck that is not easily shaken loose.

In this book emphasis will be primarily on the global warming issue and secondly on various environmental and energy issues. The political battles involved in each of these areas will also be covered. Here we will limit ourselves on the color issue to the question: *How Green are the Gorons?* Comments on this rather profound question will be offered throughout this book, then finally answered in the concluding chapter. Clearly the Gorons have tried to hijack the whole green spectrum. But that is far too broad, with so many fine shades of green. That cannot be permitted to happen. A much thinner slice of this spectrum must be selected. However, the exact definition of that slice will be left for the final chapter.

1.4 Book Organization

This sequel to my first book has eight sections.

• **Section 1**
  Introduction.

• **Section 2**
  This reviews Albert Gores efforts to con the public on the climate change issue via an ultra simplistic analysis of Earth's temperature. This is followed by a conceptual Temperature *Model* illustrating the complexity of temperature analysis. Book reviews of his two books are provided

• **Section 3**
  This initiates the coverage of key debates on global warming science, starting from the Houston debate previously covered. Two other debates are covered: one on the radiation phenomena, and one on the validity and utility of the huge computer simulations used in climate analysis.

• **Section 4**
  This section covers developments in Florida. The first chapter covers their voting problems, past and present. Other issues include the hurricane situation and the sea level rise situation.

• **Section 5**
  This section briefly revisits the energy situation in California and extends that to include their climate change initiatives. It also salutes a California pioneer in the environmental and energy areas.

• **Section 6**
  This expands coverage on energy with a major situation review on all energy sources. It starts with a major analogy that compares our energy situation to that of a plane crash in the Sahara.

• **Section 7**

Environmental coverage includes the work by Theodore Roosevelt and Rachael Carson.

• Section 8

The final section summarizes the conclusions on why I am a skeptic on the Global Warming issue and addresses the key question on *How Green are the Gorons?*.

References and Notes

(1.1) Gore, Albert, *Earth in the Balance*, Houghton Mifflin Company, New York, NY, 1992.

(1.2) Gore, Albert, *An Inconvenient Truth*, Rodale, Inc., Emmaus, PA, May 26, 2006

(2.1) Begley Sharon, et al, *The Truth About Denial*, Newsweek, August 13th, 2007. Her fellow travelers in this piece were: Eve Conant, Sam Stein, Eleanor Clift, and Mathew Phillips.

(2.2) Economides, Michael, *Propaganda as Journalism*, Human Events, August 20, 2007.

(2.3) Westbrook, Gerald T., *More on Propaganda as Journalism, including the "P, P & P Test"*, Posted initially on The ecologicPowerhouse, September 7, 2007. A search for "P, P & P" and Newsweek will bring up comments on this essay, plus the original essay.

(3) Blyth, Myrna, *Spin Sisters - How the Women of the Media Sell Unhappiness and Liberalism to the Women of America*, St. Martins Press, New York, NY, March 2004.

(4) Mallick, Heather, A Mighty Wind blows through Republican convention, CBC, September 5, 2009. See: http://www.cbc.ca/canada/story/2008/09/05/f-vp-mallick.html.

(5) Farah, Joseph, *Al Gore and the new journalism*, WorldNetDaily, February 28, 2001. Farah also reminds his readers of the many comparisons made between the statements of Al Gore Jr and those of Ted Kaczynski, the Unabomber, including the "now-famous Gore-Unabomber Quiz".

(6) McGurn, William, *Pelosi's Indefensible Bill*, The Wall Street Journal, February, 10, 2009.

(7) Espo, David, *Pelosi Says She Would Drain GOP 'Swamp'*, The Washington Post, October 6, 2006.

(8) See Reference 1, Preface. In Section 12—on The Kerry Enigma and an Introduction to the Gorons in the Senate—I wrote: "And out of that primordial ooze via the wonders of fantastic fermentation or whatever, had emerged many a new Goron. For those who think that I am jesting here, for those who think what I am suggesting is impossible, I would refer you to a recent news report where 'workers in knee-high rubber boots slosh in the buildings two vast reflecting pools, vacuuming up great green gobs of goo.'" OK, so this reference is really for the state capital of Hawaii, but it does set some sort of precedent. And it may just tell one the origins of many of the senate Gorons.

(9) *Is the House Swamp Drained Yet?*, Editorial, The New York Times, April 18, 2009.

(10) Krauthammer, Charles, *The Fierce Urgency of Pork*, The Washington Post, February 6, 2009.

(11) Walsh, Bryan, 'WE' Climate Campaign: Glossy, But Will It Work, Time, September 1, 2008.

(12) Nancy Pelosi | Issues The Environment. See http://speaker.gov/issues?id=0009.

(13) Editorial, *The first step: House pasage of energy bill marks new U. S. willingness to fight climate change*, Houston Chronicle, July 1, 2009.

(14) Will, George, *The President is CPR for the GOP*, Newsweek, September 21, 2009.

(15) Klein, Ezra, *Nancy Pelosi on Health-Care Reform*, The Washington Post, July 22, 2009.

(16) Hot Air Blog, *Krauthammer: Van Jones Truther allegations "devastating"*, September 4, 2009.

(17) Brayton, Ed, Scienceblog - *Frum vs Horowitz: Dispatches from the Culture Wars*, September 16, 2009

(18) *House may pass healthcare reform bill without voting on it Part 2*. See: http://americaswatchtower.com/20-10/03/16/house-maypass-healthcare-reform-bill-without...

(19) *Pelosi: We Must Pass The Health Care Bill So That We Can See What's In It*. See: http://politifi.com/news/Pelosi-We-Must-Pass-The-Health-Care-Bill-So-That-We.., March 9, 2010.

(20) Hitt, G. & Adamy, J., *House Passes Historic Health Bill*, Wall Street Journal, March 22, 2010.

(21) Adamy, Janet and Meckler, Laura, *Vote by Vote, a Troubled Bill Was Revived*, Wall Street Journal, March 22, 2010.

(22) Strassel, Kimberley A., *Inside the Pelosi Sausage Factory*, Wall Street Journal, March 22, 2010.

(23) *The Doctors of the House*, Editorial, Wall Street Journal, March 22, 2010.

(24) For the movie *Sunset Boulevard (1950)*, see http://www.fimsite.org/suns.html. This starred Gloria Swanson who played, somewhat autobiographically, Norma Desmond, "a deluded, tragic, ambitious actress whose career declined with the coming of the *talkies*." The male lead was played by William Holden, as "a B-movie hack screenwriter/narrator." Holden—due to a combination of circumstances, ended up in her ancient and dilapidated mansion—was soon "showered with bribes (clothes, money, flattery and other gifts) and was quickly spoiled and ensnared in her web of delusion - - -."

2. Everything you wanted to know about Albert Gore, but were afraid to ask.

2.1 The Head Goron's Simplistic Model of the Earth's Temperature

As noted in the Introduction my first book, *'Acid Rains'* [1], was triggered by the writings of Albert Gore Jr., specifically his first book and the comments he made there on global warming . Also noted in the Introduction was the conviction that Albert Gore Jr. is the chief propagandist for the Goron community and all his *fellow travellers*. The HG served as a journalist during the Vietnam War. Not only did the HG develop an interest in journalism during this war, but also in propaganda.

In Chapter 2 of my initial book, Figure 1 was presented. This chart is similar to the chart used by the HG in his lectures on global warming. Several times over the past two decades[2, 3a, 3bb] the HG has strived to couple the Earth's average temperature with the atmospheric concentration of CO_2. He will typically show a graph of **temperature** and **atmospheric** CO_2 **concentration**, plotted vs time, over a span of 160 thousand years (160 KYs). The data for this graph comes from chemical analysis of two mile deep ice core samples from Antarctica and Greenland, which is valid scientific data. However, this is a perfect example of Gore gaining some validity by *cherry picking* valid scientific data, simplifying it, extrapolating it, and misrepresenting the situation until it literally has no meaning. Indeed, the HG has defined the data behind this chart: **as the most compelling evidence of a correlation between carbon dioxide in the atmosphere and the atmospheric temperature.**

The graph for each variable is of a similar shape, and appear to move somewhat in parallel, over this period. Based on such a chart the HG would make the following claims:

· Carbon dioxide concentration and temperature have moved in **lockstep** over this period.

· Since they have moved **in tandem** over 160 KYs, it would be irresponsible to assume they will not continue to move together in the future.

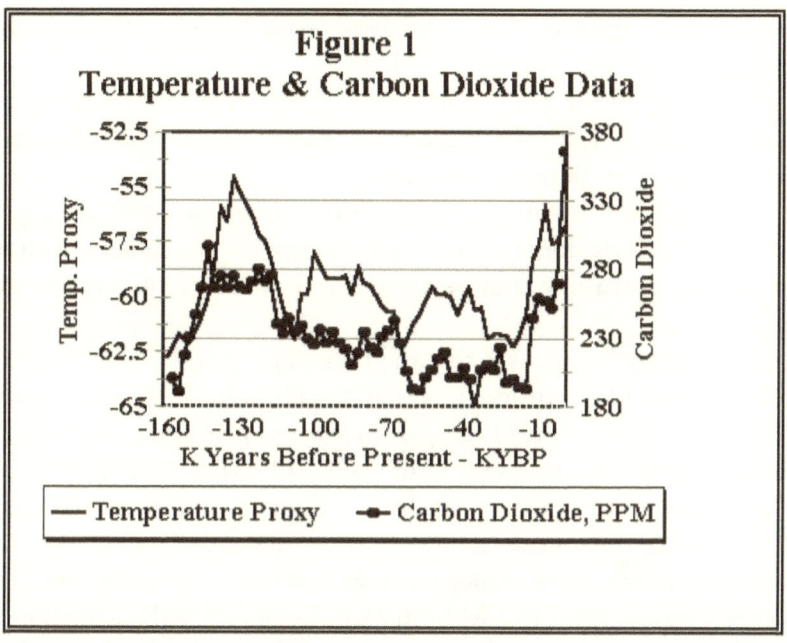

Figure 1
Temperature & Carbon Dioxide Data

However, one must object to the use of the words **lockstep** and **in tandem**, and the phrase: **continue to move together** . These are far too strong for a simple visual correlation. The implication is also made that the change in CO_2 concentration precedes the change in temperature and hence is the cause for the change in temperature. However, it is **absolutely impossible** to tell this from such a graph. Indeed one could just as easily claim the reverse relationship was proven by this chart. Yet based on this graph the HG is making a simple, visual correlation and basing all of his conclusions on that eye-balled relationship. Not what one would call a high horse-power statistical analysis.

Finally, note that the time span of 160 KYs is displayed over a space little more than two inches in width in his first book. The **data on this graph is highly compressed**. This means, for example, that the history of 20 KYs—from the time of the Last Ice Age to the present— would be shown over a space of only a quarter of an inch. The basic time unit, for the data is about 2,500 years per point. Hence one tic on this graph—about 1/32nd of an inch in width—would represent the last 2,500 years, taking us back to 500 BC.

Others have also critiqued the HG's interpretation of this data. For example, Idso[3b] reported that the scientists who developed this data, from deep antarctic ice cores, had noted that:

· In a warming era (in going from cool to warm or glacial to inter-glacial conditions) changes in carbon dioxide concentration **can not be shown to precede** changes in temperature.

· In a cooling era (in going from warm to cool or inter-glacial to glacial conditions) not only does the above statement still hold, but one **can now show that temperature change always occurs first**, followed by carbon dioxide change.

One might ask: how is it that the HG's *most compelling evidence* is so readily refuted? Idso's answer was that the HG's basic premise is simply invalid. Reason: **our planets temperature is not primarily controlled by carbon dioxide**. This is discussed in detail in Section 2.2 below.

The behavior of the two variables in Figure 1, represent the change in each variable due to the changes in other major variables, particularly the orbital parameters of planet Earth as it travels around the sun. These variables were identified by a Milutin Milankovitch, a Serbian astronomer/climatologist, from his analysis over the period 1920 to 1941[4]. But it was not until 1969 that this theory was translated to English. It has been updated and fine-tuned since then. The basic tenets of this work are that there are three attributes in the Earth's travels around the sun, that have long term impact on our climate. These attributes are:

· a 100 KY cycle in the orbital eccentricity;
· a tilt of the planet with a 41 KY cycle; and
· a wobble of the axis with 19-23 KY cycles.

While unnoticed over any individual's lifetime, these attributes—what this writer calls the Earth Motion Anomalies (EMAs)—exhibit major variation over the 160 KY period. For example the tilt of the Earth would vary over a range of about 22.2 to 24.5 degrees, with a current value of 23.5 degrees. And the eccentricity of the orbit would lead to Earth to sun distance varying:

· at today's eccentricity—of 0.0167—from 91.4 to 94.5 M miles;
· at maximum historical eccentricity—of 0.0607—from 87.3 to 98.6 M miles.

The Milankovitch Theory is widely accepted as the primary force behind the climate changes that moved the variables shown in Figure 1. Indeed these forces are the primary cause of the Last Ice Age. The changes in the EMAs cause variation in the solar insolation, the energy that hits Earth. This variation will occur for each latitude and season.

· If the solar insolation very gradually drops, due to changes in the EMAs, this will be followed by an equally very gradual temperature drop.

· Colder temperatures will lead to reduced carbon dioxide concentration in the atmosphere, as more carbon dioxide is taken up by the oceans.

· With colder temperatures more water will be locked up in ice fields.

· With more snow and ice cover, the average albedo of planet Earth will very gradually increase thus reflecting more energy.

· With more water molecules locked up as ice, the sea level will very slowly drop – over 400 feet at the peak of the Last Ice Age.

· The planet becomes drier. Humidity very gradually drops. Dust aerosols (DASLs) increase. They have been reported to have been as much as 100 times as great as to-days level.

· DASLs occurrence in the polar areas indicate that air circulation was much higher than today.

These increases in the levels of natural aerosols, and the reduction in carbon dioxide concentrations will enter to play re-enforcing roles in the ever so gradual cooling of our planet.

All of the above forces created the period known as *The Last Glacial*, a period all most unholy in its nature. Cooling, warming and more cooling occurred over and over again, with a multiplicity of relatively rapid cooling events, from 116 to 13 KY ago, the end of the Last Ice Age.

These orbital phenomena will impact climate on 100 KY and 10 KY time scales and possibly down to the 1 KY (1000 year) time span. However it is unlikely they would impact our climate over the century, decade and annual time scales that are important in the current debate on global warming. They are included here to illustrate the wide span of natural climate variation and also to help clarify the interpretation of the data in Figure 1 that has been so misrepresented by the HG.

References and Notes

(1) The following references on Albert Gore Jr. were taken from Chapter 2 of my initial book. See Reference 1, in the Preface.

(1a) Terzian, P., *Bland Ambition*, The American Spectator, August 1999. This essay is based on the book - *Gore: A Political Life* by Bob Zelnick, a former ABC News correspondent. Terzian credits Zelnick's reporting as enabling him to see Gore with some clarity and to help him to penetrate the mystery of this politician.

(1b) See *Environmental Scientist Dossier: Albert Gore, Jr.*, by The National Center for Public Policy Research, March 21 1996.

(1c) Wolfsan, Adam, *Apocalypse Gore*, Copyright 1999, National Review, March 8 1999, as reprinted with permission on www.junkscience.com/feb99/apocgore.htm.

(1d) See *Newsmakers Albert Gore Jr.*, ABCNews, 2000, quoting Vanity Fair magazine, March 1988.

(1e) Rubman, G., *Hypocrite Gore Should Practice What He Preaches*. See: www-tech.mit.edu/V116/N27/gore.27c.html, June 7 1996.

(2) Gore, Albert, *Earth in the Balance: Ecology and the Human Spirit*, Houhton Mifflin Company, New York, NY, 1992.

(3a) Gore Jr., Al, *To Skeptics on Global Warming ...* , New York Times, April 22 1990.

(3b) Idso, S., *Carbon Dioxide Warming is Good for the Planet*, New York Times, May 7 1990.

(4) Berger, A., Introduction to the Milankovitch Theory of Climate", Review of Geophysics, 26, November 1988.

2.2 Alternative Temperature Models

It was noted in the Introduction that one of the definitions of a Goron is one who embraces simplistic science as the answer to complex problems. Here, the HG's analysis of temperature is the textbook example of such behavior.

The subject of temperature analysis is a most complex story. As a means to show just how complex it is, a technique used by economists will be employed [1]. Frequently economists express, at least initially, the relationship(s) they think exist between variables as a functional specification.

Details on such an approach, with perhaps a total of 100 or so variables considered, are shown in the References and Notes section for NOTE 3.

This approach—namely the econometric model specification approach—highlights the challenge involved in the analysis of the Earth's temperature and in modeling the climate. Surely, no simple statistical model—specifically a linear regression model, sometimes called a least squares model—could contain all these variables.

Indeed the complexity involved rapidly leads to the abandonment of statistical models and the movement to the use of huge computer, simulation models. These are known as the so-called general circulation models (GCMs). While there has been some success with the GCMs, there also have been significant failures. The GCMs will be covered in Chapter 3.3.

In summary the subject of temperature analysis is a most complex story. In addition to the material covered in Chapter 3.3, comments will be made on this subject throughout this book and on the so-called hockey stick profile, another high profile and highly controversial temperature model.

References and Notes.

(1) Parts of this section were first published by this writer as - *Global Warming: Are Society's Attitudes and Actions Based on an Over Simplistic View of a Highly Complex System*, United States Association for Energy Economics, News Letter, December, 1997.
(2) Lockwood, M., et al, A Doubling of the Sun's Coronal Magnetic Field during the Last !00 Years, Nature, June 3, 1999.

2.3 Additional Comments on Earth in the Balance.

Besides the problems noted above two additional areas need amplification.

• The Gorons Graph
First, there was another major problem with the HG's CO_2 graph in his book. That problem deals with a single point, namely the last point shown (not shown in § 2.1, Figure 1). The level of CO_2 varied over the 160 KY interval from about 295 ppm, down to about 180, then moved back up to about 270. Then, on the HG's graph, it *sky rocketed* to 600 ppm and beyond. In the text he does talk of the current level of CO_2 is at 355 ppm, but one cannot pick that up off his graph. And this 600 + ppm is not part of the scientific record. Rather it is a scenario projection., When one includes inputs on a graph of different type, source or definition these must be clearly noted. This was not done. Further, the inclusion of that one point brings the graph height to over seven inches, on a graph that would only be two to three inches otherwise. In short this is pure distortion and mis-representation.

• Gore's Disturbing Writing
On a second subject, some of the writings by Gore are quite disturbing and brand our former VP as an extremist. Several quotes are examined.
· He equates the dangers to the environment to the dangers of nuclear war. Can this be a truly sound judgement of the state of the environment after

years of the EPA, after years of automobile fuel and exhaust improvements and after billions of dollars of environmental control investments in all our utility and manufacturing industries?

· He argues that "the evidence of an ecological Kristallnacht is as clear as the sound of glass shattering in Berlin." Can this be viewed as a precise and sound diagnosis of the ecological situation?

· He argues that Americans' use of natural resources is equivalent to Nazism.

· He charges that our civilization must be considered in some basic way dysfunctional. "In psychological terms, our rapid and aggressive expansion into what remains of the wildness of the earth represents an effort to plunder from outside civilization what we cannot find inside."

· He argues that our society's embrace of what he calls consumptionism, resembles Nazi Germany society's embrace of totalitarianism.

· Next he claims we live in an "inauthentic world of our own making. "Life can be easy, we assure ourselves. "We need not suffer the heat or the cold; we need not sow or reap or hunt and gather. "We can heal the sick, fly through the air, light up the darkness, and be entertained in our living room by orchestras and clowns whenever we like." This false world was created by people to distract people from their psychic pain. He asserts that the world of leisure, air conditioning, industrial agriculture, modern medicine, and home entertainment is not good in itself. "It is but a fleeting sideshow." Finally he claims that only by somehow awakening from such in-authenticity will the cycle of psychic pain and environmental plunder be broken.

It is not quite clear how the HG—who grew up in a hotel in Washington—earned credentials to decide what is authentic in life and what is not. Or who is authentic and who is not.

• The **Unabomber or Al Gore?**

As noted above Gore has a strange streak of extremism in his makeup. He does not seem to be happy with our society. He equates environmental activists to resistance fighters. But does not this group of activists include eco-saboteurs? Clearly Gore does not limit his concern to just the environment, but declares that we are in a midst of political, informational, inner-spiritual and deep philosophical crisis. He is credited with being the sole author of his book and to have put his heart and soul into it. Surely he believes in his book and all its comparisons of our society with Nazi Germany's or to the former USSR's. All of the above quotes and comments have earned him the extremist tag. No where has this been better exposed than in a 1995 column by Tony Snow[1]. In this essay he contrasts statements by Gore to those of the Unabomber. Snow contrasts statements by Gore to those of the Unabomber. Snow noted that the vision advocated by the Unabomber sounds much like that stated in Gore's 1992 manifesto. The infamous Gore/Unabomber quiz is listed below[2]. The difference between the Unabomber and Gore is that Gore wants to achieve his goals via highly intrusive regulations and massive government bureaucracies, while the Unabomber would achieve this through mail bombs

• **Conclusions**

I could go on, but I think the above is enough. However, for those readers who would like additional inputs on this subject I would recommend the book Environmental Gore [3], edited by a John Baden. The first section, *Civilization in the Balance*, is a particularly good start for this review.

Did Al Gore say it? Or was it the Unabomber? [2]

1. "The twentieth century has not been kind to the constant human striving for a sense of purpose in life. Two world wars, the Holocaust, the invention of nuclear weapons, and now the global environmental crises have led many of us to wonder if survival - much less enlightened, joyous, and hopeful living - is possible. We retreat into the seductive tools and technologies of industrial civilization, but that only creates new problems as we become increasingly isolated from one another and disconnected from our roots." (G).

2. "Again, we must not forget the lessons of World War II. The Resistance slowed the advance of fascism and scored important victories, but fascism continued its relentless march to domination until the rest of the world finally awoke and made the difference and made the defeat of fascism its central organizing principle from 1941 through '45." (G).

3. "It is not necessary for the sake of nature to set up some chimerical utopia or any new kind of social order. Nature takes care of itself: It was a spontaneous creation that existed long before any human society, and for countless centuries, many different kinds of human societies coexisted with nature without doing it an excessive amount of damage. Only with

the Ind. Revolution did the effect of human society on nature become really devastating." (U).

4. "Modern industrial civilization, as presently organized, is colliding violently with our planet's ecological system. The ferocity of its assault on the earth is breathtaking, and the horrific consequences are occurring so quickly as to defy our capacity to recognize them, comprehend their global implications, and organize an appropriate and timely response. Isolated pockets of resistance fighters who have experienced this juggernaut at first hand have begun to fight back in inspiring but, in the final analysis, woefully inadequate ways." (G).

5. "Among the abnormal conditions present in modern industrial society are excessive density of population, isolation of man from nature, excessive rapidity of social change and the breakdown of natural small-scale communities such as the extended family, the village or the tribe." (U).

6. " All pre-industrial societies were predominantly rural. The Industrial Revolution vastly increased the size of cities and the proportion of the population that lives in them, and modern agricultural technology has made it possible for the Earth to support a far denser population than it ever did before." (U).

7. "The positive ideal that is proposed is Nature. That is, wild nature: those aspects of the functioning of the Earth and its living things that are independent of human management and free of human interference and control." (U).

8. "Any child born into the hugely consumptionist way of life so common in the industrial world will have an impact that is, on average, many times more destructive than that of a child born in the developing world." (G).

9. "And tragically, since the onset of the scientific and technological revolution, it has become all too easy for ultrarational minds to create an elaborate edifice of clockwork efficiency capable of nightmarish cruelty on an industrial scale. The atrocities of Hitler and Stalin, and the mechanical sins of all who helped them, might have been inconceivable except for the separation of facts from values and knowledge from morality." (G).

10. "The modern individual on the other hand is threatened by many things against which he is helpless: nuclear accidents, carcinogens in food, environmental pollution, war, increasing taxes, invasion of his privacy by large organizations, and nationwide social or economic phenomena that may disrupt his way of life." (U).

11. "Industrial society seems likely to be entering a period of severe stress, due in part to problems of human behavior and in part to economic and environmental problems." (U).

12. "What does it say about our culture that personality is now considered a technology, a tool of the trade, not only in politics but in business and the professions? Has everyone been forced to become an actor? In sixteenth century England, actors were not allowed to be buried in the same cemeteries as 'God-fearing folk,' because anyone willing to manipulate his personality for the sake of artifice, even to entertain, was considered spiritually suspect." (G).

2.4 An Inconvenient Reincarnation – a Critique of An Inconvenient Truth.

• Background

Our former VP has received rather incredible coverage on this, his second book[1.1] and movie[1.2]. Clearly most of this coverage has provided him with many accolades and awards, including an Oscar and a share in a Nobel Peace Prize. There have also been a rather incredible number of critiques on this book[2].

First of all there are many pictures of Albert Gore in this book in some natural setting, which is fine. If he had not been born in to a political family he might have been content to have been a forest ranger or a national park guide.

Next there are many pictures of Albert Gore in this book in some family setting, which is also fine. I surely do not know his family, but the pictures—of his father, sister, wife, children and grandchildren—suggest to this reader that this family, at one time, was in the same league as JFK's.

However, I approached *AIT* believing it would be loaded with propaganda. It was. I see this:

· in his selection and coloring of many pictures;

· in his analysis of selected temperature trends; and

· in other concerns in the cryosphere, hydrosphere and the biosphere. Only a few will be noted.

• Use of Colored Pictures

Gore's current book is full of high quality colored pictures, graphs and maps, just the opposite to his first book. However, this is part of his propaganda pitch. Three examples follow.

(1) <u>Colored Pictures on Astronomical and Astrophysical Forces</u>. About 25 pages are devoted to superb astronomical and astrophysical pictures. With such a display one might think that Gore was the leading advocate of orbital, solar and astrophysical influences on our climate, and on global warming (GW). However, the exact opposite is the reality.

· Orbital parameters: he makes no mention of the Earth's eccentricity, tilt and wobble, factors that were, most likely, the primary forces behind the Last Ice Age, and earlier glaciations.

· He makes no mention of the changes in our sun such as: sun spot activity, solar flares, the solar wind and the solar magnetic activity[3].

· Finally, he makes no mention of the possible interaction of cosmic rays on cloud formation[4].

Gore has completely turned his back on these factors. indeed on the whole astrophysical sciences and there role in temperature change and on our climate in general, but not on such pictures.

(2) <u>Colored Pictures on Greenland Ice-sheet Collapse and Resultant Sea Level Rise</u>. In another example, Gore shows three Greenland pictures/maps with snow–ice cover shown in white and the surface meltwater in very bright red. Perhaps I wouldn't be so upset if he had shown the meltwater in a pastel grey or blue and not red as that color surely distorts the situation.

First off, Gore gives zero definition as to what is meant by meltwater. He does not define what fraction of this surface area would qualify as "melting." Nor does he define the depth of meltwater: tens of meters; tens of centimeters, tens of millimeters or what? Yet such definition is fundamental to an analysis of the state of this ice-sheet. Next he gives only the year, but no inputs on the date during that year that each picture/map represents. Surely the degree of melting shown must be significantly different if the pictures were taken in late August versus early April.

A fourth map—which was not shown—would be a similar projection of what the Vikings would have observed 1000 years ago. There must have been surface melting then, when they settled.

One should note that his three picture/maps represent only three points in a very long history of a very complex subject. Surely a table of say fifty years of history, with two or three points per year—giving the day/month/year and the percent of Greenland area meeting or exceeding some agreed definition of melting—would be a much more valid scientific starting point to understand what is going on in Greenland. Are any trends shown? Is there indication of any cycles? However, such a table would surely not have the impact of his three dramatic pictures.

The conclusion, that the uninitiated will draw, is one of imminent and terrifying collapse of this ice sheet and subsequent devastating sea level rise. However, Gore does not give the average reader any sense of the chances that such an event could happen. Yet many references define this as an incredibly low probability outcome. Additional comments on this subject can be found in §4.3.

(3) Colored Pictures on Hurricanes. The same technique—as that used on the Greenland picture/maps—is used on hurricanes with 20 pages of pictures on hurricanes, many in shades of red. The implied conclusion, again, is one of imminent catastrophic climate change (CCC). Michael Crichton, whose book *State of Fear*, addresses CCC, noted: the "politico-legal-media complex must keep the populace in a continuous state of fear." Gore's new book surely practices this objective. More on hurricanes follow.

• Temperature Profiles

A major objective of Gore's book is to reinforce the conviction that there is an unshakeable correlation between CO_2 concentration in the atmosphere and global temperature. The new book has five temperature charts versus two in his first book. Three of these will be reviewed.

(1) Temperature Profiles: 650,000 years (650 KYs) ago up to the present — This is an extension of the chart used in his first book. It is now 9 inches wide vs 2¼ in his first book. However, there is an error in the time scale for the 300 KY to 600 KY period. It is estimated that one centimeter on this graph covers 20 KYs. Hence, if each tick is one mm in width, that tick would represent about 2 KYs of history. As noted in my review of his first book the graphs exhibit a *saw-tooth* profile. And the same conclusions apply: it is impossible to tell from this graph that changes in

CO_2 lead to changes in temperature. Indeed, the causality mostly runs from temperature change to CO_2 change[9].

(2) <u>Temperature Profiles: 1,000 AD up to the present</u> —. Gore's horizontal graph shows the temperature profile as ~ flat for 950 years, than taking off for the last 50 years. This is the *hockey stick*. First published in 1998, the *warmers* have strived to get this *history revision* accepted. However, the fight over this *stick* is not over. Like the 100 year war, it goes on and on. The volume of references on this battle is huge.

The older version of this temperature profile consisted of a Medieval Warming Period (MWP), the Little Ice Age (LIA) and the current warming period.. The warming periods on Gore's graph are shown in red, with about 96% of this in the last 50 plus years. In his first book he reports on the so-called Medieval Warming Epoch a period from about 900 to 1300AD. He noted that "it clearly seems to have been a shift in the global climate pattern." He also noted this climate shift was of such a magnitude that it was the reason they were able to go from Scandinavia to Iceland to Greenland and ultimately to North America. In contrast this Epoch was barely flagged on one of his graphs of the *hockey stick*, as a tiny blip around 1360AD.

The LIA received considerably more coverage in his first book. He noted that temperatures began to drop at the start of the fourteenth century, and he described the LIA as "one of the most important and well-documented climate fluctuations", a period from 1550 to 1850 AD. This period is not mentioned in his new book or noted on the *hockey stick* graphs.

(3) <u>Temperature Profiles: 1860 to 2005</u> —_The chart from the Intergovernmental Panel on Climate Change (IPCC) shows the following four temperature trends.

Trend	Interval	Departures from the 1961 – 1990 average	Change over the Indicated interval	Comments
I	1860 - 1905	- 0.42 to - 0.42	0.0	Flat
II	1905 - 1945	- 0.42 to + 0.04	+ 0.46	Up
III	1945 - 1975	+ 0.04 to – 0.12	- 0.16	Down
IV	1975 - 2005	- 0.12 to + 0.65	+ 0.73	Up

Temperature Trends as Estimated from the IPCC Chart º C

However, Gore neglects to note the idiosyncracies in this chart. These can be seen in his chart, but are more obvious in other versions of this chart

— yes, there are other versions of this chart that use different databases and/or data massaging techniques. The two key idiosyncracies are:

· there is **significant warming over Trend II** where CO_2 output is just barely beginning to rise,

· and there is **cooling over Trend III** where CO_2 growth really takes off after 1945. If changes in CO_2 are truly the key force in changes in temperature then one should not have seen the warming in Trend II and the cooling in Trend III.

There is another issue with this graph that relates to data quality. Such subjects as station inclusion, precise equipment location, equipment changes, paint changes, site changes, station management, data management, adjustments used for the Urban Heat Island effect are all of interest.

Recently a Dr. Richard Lindzen—perhaps one of the most significant skeptics on climate change and the leading academician in this field—highlighted some of these problems on the three major weather station databases[10]. Some of his comments follow.

· "While there is agreement regarding the historical data, that does not mean that the resulting observations are very solid."

· Interestingly the IPCC "record is essentially flat since 1995." To support this claim Lindzen re-plots the data for 1995 to 2005 as a bar chart, with each bar about 8 mm wide versus 1 mm in Gores book. The difference in the visual presentation of this data is remarkable. Further, if the 11 points are divided into samples of five and six values, the averages of these departures (from the 1961 to 1990 average) are 0.39 and 0.44 °C. However, the difference between these two averages is statistically insignificant. Statistically, one cannot see a difference.

· Two of the records indicate 2005 was a record breaker, "by a statistically insignificant amount[10]."

• **The Cryosphere**

(1) <u>Small Glaciers</u> – Gore makes no mention of the complexity of this subject and that glaciers have been advancing and retreating since the last ice-age. Examples follow.

· Alaskan glacier[11]. The Hubbard Glacier has been advancing for more than 100 years.

· Andean glaciers[12], have advanced four times over 1250-1850, each during a solar minimum.

· Himalayan glaciers[13]. Professor Zhang Wenjing—with the Chengdu Institute of Mountain Hazards and Environment, under the Chinese

Academy of Sciences—discounted forecasts that Himalayan glaciers could disappear in 50 years. "These predictions may be excessively pessimistic."
· Mt Kilimanjaro glaciers[14., 15]. Gore claims the near snow free peak is due to GW. He, along with Senators Hillary Clinton and John McCain ignore inputs that point to:
(a) no temperature trends in this area from at least 1948 to 2005;
(b) drops in humidity and atmospheric moisture content and reduced rain/snow, since 1880; and
(c) the vertical edges of these mountain plateau glaciers have become exposed to solar radiation.

While all ice bodies on Kilimanjaro retreated from 1912 to 2003, the highest retreat rate occurred in the first part of the 20th century. The rate from 1989 to 2003 was the lowest of all intervals. Apparently the exposure of the vertical margin dominates the retreat of these glaciers. The rapid recession in the first part of 20th century shows they were drastically out of equilibrium then, due to a relict climate change that occurred in the late 19th century.
· New Zealand glaciers[16]. The authors cited a study a study on four NZ glaciers in Mount Cook National Park. This study was primarily focused on the assessment of the presence or absence of the Little Ice Age in NZ (it was present) over the interval from 1725 to 1895. However, it also provides some useful input on NZ glaciers. The scientists reported that "after a slow, but constant retreat during the second half of the 18th century and the first half of the 19th century, the glaciers experienced major re-advances during the second half of the 19th century (around 1869 to 1890/95)." (2)

Ice Sheets

· Antarctic Ice-Sheets I. On a large map of Antarctica in his book Gore mis-labels the West Antarctic Ice-Sheet (WAIS) as an ice-shelf. In the text he again refers to the East Antarctic Ice Sheet (EAIS) as an ice-shelf, and the WAIS, twice as a shelf. Now this may sound like a trivial error to those unfamiliar with this subject. Some might even call this a *typo* except it happened four times. More importantly there is a huge difference between an ice-sheet and an ice-shelf. An ice-sheet is grounded on hard rock. An ice-shelf floats on water, but is tied to the continent. A third category is sea-ice that floats free. The ice-sheet category is the only one that contributes to a SLR on melting. However, Gore compounds this error by stating that the melting of an ice-shelf will add to the SLR.

· Antarctic Ice-Sheets II. A huge number of references focus on whether the EAIS and/or the WAIS are growing or shrinking. A 2005 reference provides a sample of this traffic[17]. It also alerts its readers that the vast majority of GW reports on Antarctica are focused on the Antarctic Peninsula. This region represents about 2% of the total. More important it may be more akin to the forces that control the conditions on southern South America. In short conditions on or around this peninsula are not representative of the continent as a whole.

The above sample listed eight reports that claimed this continent was shrinking, and seven to it growing. One of these reported increased snowfall and accumulation from 1992 to 2003, leading to a mass gain of 45 ± 7 billion tons/year. This is tying up enough water to lower the SLR by ~0.12 mm/year. Another author reported that at: "the rate of retreat observed in the late 1990s, the WAIS should disappear in about 7000 years." One would not call this outlook as imminent.

· Greenland Ice Sheet I. Gore shows a sketch in his new book of a profile of an ice sheet, with major water flow to its base. He depicts a moulin—a massive torrent of fresh meltwater—tunneling straight down to the bedrock below the ice. This is a subject that I believe needs to be treated very carefully. I think the conditions of when and where this phenomenon occurs needs to be defined. As these ice sheets can be two miles thick, it is hard to believe that this water remains fluid for very long. In any event this issue is listed under the speculative category. The writer stated that the glacial flow associated with moulins is tiny.

· Greenland Ice–Sheet II. Many reports show snow has increased in the interior and decreased on the edges. A recent report noted it can change by **± several meters** over 20-30 years and concluded it might have to be studied for decades to see if it is thinning[18].

• The Hydrosphere

(1) Sea Level. In AIT Gore talks of an 18 to 20 foot SLR. He shows twelve sobering pages of the impact on Florida, San Francisco Bay, New York and other areas of such a rise. But he totally ignores the 2001 forecasts by the UN (via the IPCC) of a rise of only 0.3 to 2.9 feet by 2100. Hence, the 20 foot number in Gore's book looks pretty ridiculous, unless one is into propaganda.

However, the conclusions that the uninitiated will draw from his pictures is one of imminent and terrifying collapse of the Greenland ice sheet and subsequent devastating sea level rise. Gore does not give the average reader any sense of the chances that such an event could happen. Yet there are many references that define this as an incredibly low probability outcome.

(2) Hurricanes. Gore claims warm water is the cause. However, he ignores inputs—from such scientists as Dr. James O'Brien, the Florida State Climatologist and the Distinguished Professor of Meteorology and Oceanography at Florida State University—that the ocean has not warmed in the hurricane formation part of the Atlantic. From 5 °N to 20 °N and from Africa over to the U. S. The ocean has cooled. O'Brien also sees a 15 year building trend and a 15 year decaying trend. Gore ignores this approximate 30 year cycle on hurricanes.

(3) The Global Ocean conveyor Belt. This Belt is sometimes called the ThermoHaline Circulation (THC). This system is primarily driven by temperature and salinity differences. In contrast, the Gulf Stream is a wind driven system influenced by the Earth's rotation and the lunar tides. Gore reported that this THC "pump" plays a crucial role in powering this circulation, but global warming could disrupt this phenomenon. Gore also reported around 10,000 years ago the "pump" began to turn itself off, and the Gulf Stream virtually stopped. Lindzen recently noted that to shut down the Gulf Stream you'd have to stop the rotation of the Earth or shut off the wind [9, 19].

• The Biosphere

In spite of hundreds, perhaps even thousands of reports on enhanced crop and forest growth from the increased atmospheric CO_2, Gore barely mentions this photosynthesis bounty, and then only as one of his misconceptions. I will limit my comments to one reference. This CO_2 bounty is credited, at least in part, with the so-called Green Revolution, with the higher levels of CO_2 being very favorable for food production and food security[20]. The interested reader is referred to the work of Dr. Sylvan Wittwer in general.

• Conclusions.

For those readers who would like additional inputs on this subject I would recommend the writings by Marlo Lewis Jr., especially his *Al Gore's An Inconvenient Truth*.[9] In this paper Lewis lists 93 concerns:

· one-sided - 26;
· misleading - 15;
· exaggerated - 8;
· speculative - 26; and

· wrong - 18.

As a means to bring this essay to a close I will use some of the thinking and convictions[21] from Bjørn Lomborg (BL), a Danish, environmental/statistician. Lomborg brilliantly "demonstrates the ways in which professional environmentalists play fast and loose with the truth." In short some "Environmentalists manipulate data, so as to create a false picture about the state of the world. There is *much food for thought* here.

I believe exactly the same situation exists on the global warming issue. I have described his first book as "An excellent example of alarmism, inappropriate interpretation and outright misrepresentation ." His second book can be described in the same manner, but even more strongly.

However, starting in November of 2009 we have had the so-called *Climate-gate* scandal, where thousands of emails between *warmers* were either hacked or leaked to the internet. The coverage by key news outlets is highlighted below by their own titles or sub-titles.

· NewsBusters[22] "NY Times Tackles Damming GW Emails, But Reveals Own Hypocrisy."

· The Washington Post[23] "Scientists emails deriding skeptics of warming become public."

· Wall Street Journal[24] "The emails that reveal an effort to hide thetruth about climate science."

· Spiegel Online International[25] "A Superstorm for Global Warming Research."

I have spent a part of my career as an energy economist. As such I am intimately familiar with the difficult tasks of forecasting and the demands made by those receiving such forecasts. The recipient of any forecast is most interested in the assumptions made, in the track record of the forecaster and in the overall capability claimed. Perhaps there never has been an area where the "forecasters" have claimed such an incredible level of capability as in climate change. Temperature, heat waves, droughts, tornadoes, hurricanes, even epidemics are all forecasted by the warmers. One of their more incredible "forecasts" they make is that catastrophe is just around the corner and the world has to act immediately or *the meteorite will hit.*

What is needed in this field is more research and much more transparency on key studies, on data management and on analytical techniques used. Surely we will know so much more after ten to twenty years of additional research and analysis on this subject. And society will be much more capable to assess if these forecasters are truly legitimate or is there high level *snake oil salesmanship* wrapped up in the grand colored presentations such as *An Inconvenient Truth*.

References and Notes

(1.1) Gore Jr., Albert, *An Inconvenient Truth*, Rodale, Inc., Emmaus, PA, May 26, 2006

(1.2) Gore Jr., Albert, Movie *An Inconvenient Truth*

(2) Part of this chapter was published by this writer as - *An Inconvenient Reincarnation - A Critique of An Inconvenient Truth*, Dialoge, U. S. Association for Energy Economics, March 2007.

(3) Hoyt, Douglas V. Et al, *The Role of the Sun in Climate Change*, Oxford University Press, 1992.

(4) Svensmark, H., et al, *Variation of cosmic ray flux and global cloud coverage ... a missing link in solar-climate relationships*, J. of Atmospheric and Solar-Terrestrial Physics, 59, 1225-1232, 1997.

(5 - 8) Not assigned.

(9) Lewis Jr., Marlo, *Al Gore's: An inconvenient Truth*, See Competitive Enterprise Institute publication *On Point*, September 28, 2006

(10) Lindzen, Richard, *Global Warming Facts – Some Relevant Figures for Current Behavior of Global Mean Surface Temperature*, See www.ecoworld.com, October 15, 2006.

(11) Trabant, D. C., et al, *Hubbard Glacier, Alaska: Growing Faster and Advancing in Spite of Global Climate Change - - - ,* USGS, January, 2003.

(12) *Global Warming: Some Inconvenient Glaciers*, published by the SEPP newsletter, The Week That Was, June 17, 2006. The writer quotes the paper: *Solar modulation of Little Ice Age climate in the tropical Andes*, Polissar et al, Proceedings of the National Academy of Sciences, 2006.

(13) *Himalayan glaciers 'safe for centuries'*, News in Science, October 18, 2006.

(14) *The Retreating Glaciers of Kilimanjaro*, September 20, 2006, Center for the Study of Carbon Dioxide and Climate Change. The authors review a paper by six glaciologists (Cullen et al), published in Geophysical Research Letters.

(15) Glacial Retreat on Kilimanjaro, March 10, 2006, Center for the Study of Carbon Dioxide and Climate Change. The authors review a paper by five scientists (Kaser et al), Int. J. of Climatology.

(16) *Glacial Behavior in New Zealand*, December 22, 2004, Center for the Study of Carbon Dioxide and Climate Change. The authors review a paper (by S. Winkler), published in *The Holocene*.

(17) A Global Warming Snow Job? See www.worldclimatereport.com, May 27, 2005

(18) *The Need for Long-Term Glacier Mass Balance Data*, October 18, 2000, Center for the Study of Carbon Dioxide and Climate Change. The authors review a paper (by R. J. Braithwaite and Y. Zhang), published in *J. of Glaciology*.

(19) Statement made by Richard Lindzen during a debate on global warming entitled: *Could Global Warming Kill us?* See Transcripts: CNN Larry King Live, January 31, 2007.

(20) Wittwer, Sylvan, *The Global Environment and Food Production.*
See www.greeningearthsociety.org, June 20, 2006.

(21) McCarthy, Michael, *The Skeptical Environmentalist by Bjørn Lomborg, A cool head in the hot air,* THE INDEPENDENT, August 31, 2001.

(22) Waters, Clay, *NY Times Tackle Damming GW Emails,* Newsbusters, November 19, 2009.

(23) Eilperin, Juliet, *Hackers steal electronic data from top climate research center,* The Washington Post, November 21, 2009.

(24) Evers, Marco et al, *Climate Catastrophe - A Superstorm for Global Warming Research,* Spiegal Online International, April 1, 2010.

3. Key Debates on the Global Warming Issue

In a sense the overall dialog between *warmers* and skeptics over the past 10 - 20 years has been a debate. However, the *warmers* would not admit to that. They would insist there has been no debate at all. In the past the existence of a debate has been down-played and skeptics have frequently been depicted as few in number, negative, on the marginal side of the science and even a bit crazy. More recently they have been depicted as deniers, calling up pictures of the Holocaust.

The warmers have insisted, and continue to insist that the science is done, and that all scientists agree on that position. Indeed the rather incredible use of the phrase "the science is done and that all scientists agree on that position" should be *food for thought* for all those who have not followed this issue intimately. In any event, if it is not a debate it surely has been and continues to be a full fledged argument. Some might call it a never-ending hissing contest. It hasn't all been fun.

These arguments have covered essentially every aspect of this issue: amount of warming; the nature of this warming - catastrophic or benign; the cause of the warming - society or natural; the temperature history; the status of glaciers; amount of sea ice; condition of the Greenland and the Antarctic ice-sheets and on and on and on..

These arguments are not simply focused on key technical questions, but on what specific reports say, and don't say; and on who is a qualified scientist, and who is not. The position that only "peer reviewed" scientific papers have any validity is stated. However, the "peer reviewed" process has become so politicized that it is essentially bankrupt.

Three debates will be covered here:
• the Houston Debate - an update on the debate covered in my initial book;
• the debates over Climate Sensitivity;
• a review of the role and status of the huge computer models used in this field.

Chapter 3.1 Debates on the Global Warming Issue - The Houston Debate

• Introduction

This 1998 debate[1] led to a surprising conclusion on the Kyoto Protocol. This debate was particularly important to me as I was there, with pen, paper and recorder. What was achieved was to see these individuals *in action* and to observe that a deep and complex gulf exists on GW.

It brought together the following group of global warming scientists and one prominent science writer.

Participant	Organization	Category
Dr. James Hurrell	Nat. Center for Atm. Research, Boulder	Proponent
Dr. Jeffrey Keihl	Nat. Center for Atm. Research, Boulder	Proponent
Dr. Stephen Schneider	Env. Biology & Global Change, Stanford	Proponent
Dr. Gerald North	Head, Climate Research Project, Texas A&M	Neutral
Dr. John Christy	U. of Alabama, Huntsville; Satellite Data Base	Skeptic
Dr. David Legates	S. Regional Climate Center, Louisiana State U.	Skeptic
Dr. Richard Lindzen	Alfred P. Sloan Professor, MIT	Skeptic
Dr. Richard Kerr	Science Magazine	Journalist

Participants in the Houston Debate on Global Warming Science.

The skeptics at this session came across as positive, brilliant, human and interesting. Some key inputs from these participants follow.

• Dr. Richard Lindzen of MIT.

Lindzen is possibly the leading academician in the climate field and perhaps the worlds most pronounced global warming skeptic, He has taught courses on such subjects as American musical comedy.

The problem of noise in the data and the noise in the overall communications on this subject was noted. Dr. Lindzen commented that most of what the public knows about global warming does not come from the scientific community, but rather from advocacy groups such as the Union of Concerned Scientists, the Sierra Club and so on. And some people from such groups distort things.

Lindzen noted that we are talking of very very small temperature changes. He suggested that natural climate variability needs a great deal more emphasis. He discussed 3 to 4 areas of natural climate variability that the large computer models do not pick up at all or do so with insufficient details or accuracy. The El Niño is the best known example of such natural climate variability.

He reported a problem today with the testing of the computer models of the climate. Today modelers use estimates of the natural climate variability, obtained from very long term runs of a model, to test the model. He feels this approach is "on pretty shaky grounds".

• **Dr. John Christy** from NASA and the University of Alabama at Huntsville.

Christy is one of the key driving forces behind the satellite based temperature data. He is also a minister, a missionary in Kenya and a marathon runner.

Christy effectively defended the satellite based temperature record over the past ~20 years. Several adjustments have been necessary. Media reports seem to present such changes in a fashion to convey that these finally resolve major differences with surface based weather station data, and as a result, we are left with the conventional wisdom that the world is warming. Christy was confident that the basic differences in temperature trends remain. These show the satellite data with very little, if any, temperature trend versus a positive trend for surface based data. His conclusion is also supported by balloon data measurements and a third source — night marine air temperatures.

Christy also expressed concern on the recent flurry of reports, on so-called extreme climate events, as evidence of global warming. As an example of this kind of hype, he cited the reports on the extreme drought in Texas over the summer of 1998.. He noted while Texas was dry this summer, the worst period by far was the 1930s. In that multi-year period, drought existed all the way from Canada down to Mexico. He noted, in contrast, this year Kansas has had bumper crops.

Christy concluded that climate is changing — it always has and always will. While a fraction of that change may seem to be coupled to human activities, no one knows how much. Dr Christy assured the students present that the current generation of climate scientists "will leave you lots of interesting problems to solve".

• **Dr. Gerald North**, from Texas A&M.

North, while not noted as a skeptic, has been one of the more realistic non-skeptics. He noted that there are *traps and minefields* all over the detection activities, and also political pressure in doing research on this subject. He felt that long-term climate simulations can help to understand the noise in the system. He introduced the radio analogy where you have a signal and static. And that is what we have with climate research. He noted you are looking for very faint signals in a very noisy system. North argued:

(i) that the solar signal is not yet detectable;

(ii) that the volcanic signal is easily detected; and

(iii) that the greenhouse gas and aerosol signals are detected, but each are large and are near canceling each other out, so that their strength estimates are likely inaccurate.

Is this status enough for use on policy analysis questions? North answered his own question with a "not sure — maybe can do some things".

• **Richard Kerr**, from Science Magazine.

Kerr opened the meeting. He noted that atmospheric concentration of CO_2 has grown at about a half a percent per year, and is now up to 365 parts per million (ppm). This is the major greenhouse gas (GHG), with other GHGs accounting for about 40 percent of the *forcing* that causes warming. Kerr noted it was difficult to forecast changes in CO_2 concentration, and reported that forecasts for these GHGs had dropped five fold from earlier levels.

Kerr reported, that over the past 130 years, an increase in ground based temperature data of 0.5 degrees centigrade had occurred, with a range of 0.3 to 0.7 degrees. He also noted that there are surprises and unexpected behavior in this field. One such area is the existence of non-linear factors in this field.

Kerr noted that about 20 years ago the computer models—the so-called General Circulation Models (GCMs)—gave a warming from 1.5 to 4.5 degrees centigrade. Kerr observed we haven't improved much on this estimated range in-spite of very major changes in the computer models used.

Kerr has followed this issue for years. Perhaps his signature accomplishment was coining the phrase *fudge factors* for flux adjustments used in the GCMs. For example he once wrote that "climate modelers have been 'cheating' for so long it's almost become respectable[3]."

The word flux is a term used to denote an energy flow in watts per square meter. There are many such flows noted in weather and climate science and inside the GCMs. For example:

(i) the solar constant at ~342 watts/m^2 as the input from the sun at the top of the atmosphere;

(ii) the greenhouse effect at ~2.7 watts/m².

The so-called fudge factors are/were used to prevent these models from drifting into rather unstable computations. Additional comments on such factors are covered in §3.2 on radiation.

While there were no specific talks devoted to policy issues, the Kyoto Treaty—sometimes referred to as the Kyoto Protocol(KP)—had a definite influence on this debate. The KP was established in December 1997 at a UN meeting held in Kyoto, only nine months prior to this debate.

• Conclusions

(1) <u>Expected Temperature increase</u>. The eight panelists were queried on the expected anthropogenic warming over the next century, given a doubling of CO_2. Two came in at 2 °C, three came in at between 1 - 2 °C, two came in at 1 °C and one came in at 0.3 °C.

(2) <u>Kyoto Treaty</u>. A second query was: would they sign the Kyoto treaty? Six of the seven scientists said no. Dr. Stephen Schneider, from Stanford voted yes. Gerald North voted no, a change from his prior position based on inputs that a fully implemented treaty would only save an estimated 0.2 °C of warming by 2100.

These inputs came from a Dr. Tom Wigley of the National Center for Atmospheric Research, a noted proponent in his own right.. He published[4] the results from his latest computer runs in 1998. His results seemed to have had a major impact on nearly all of the participants at this debate. He found the KP, if fully implemented by all involved nations by 2010,—a very very low probability of occurring—would reduce warming a trivial 0.07°C by 2050, and another trivial 0.13°C by 2100. These amounts are so minuscule as to be unmeasurable. This means societies are literally being asked to spend trillions, on a policy that we won't know is ever doing any good. Yes, the proponents expect societies to invest trillions on this issue without any hope of knowing if these investments are ever doing any good.

(3) <u>Key Questions</u>. A key question was from a student seeking guidance on what his generation should do in planning for the future. Dr. Christy suggested:

· learn how to think;

· find out why people think the way they do;

· find out where is the data they are using coming from; and

· find out what kind of agendas may be behind these sources.

Another way to state this would be as an analogy to the *signal to noise ratio* discussed earlier. Students—no indeed all of us—are besieged with thousands of messages every day from TV ads, TV programs, tele-marketers, newspapers, the Internet, political pitches and spin and so on. This writer wrote a paper about 15-20 years ago on what was termed the emerging communications revolution. While correct on the issue and direction, the incredible magnitude of this revolution was totally missed as the traffic volume anticipated , and message quality, has been totally eclipsed. We are living in a world of very low signal to noise ratio. Hence students, to become effective, need to function like *World War II radio operators striving to filter valuable intelligence out of the daily propaganda stream.* They need to learn how to process this huge data flow, assess it, deflect most of it, filter out some of the noise, store it in the back brain cells, retrieve it as needed, and compare it to new inputs from new sources. Without this capability individuals will become easy marks for the industrial, commercial, environmental, educational and political shysters.

References and Notes

(1) Parts of this report have been published at:
· eco.logic, Number 46, Spring 1999.
· International Association for Energy Economics, News Letter, 4th Quarter, 1999.
(2) This meeting was held in Houston on September 25, 1998. The seminar was sponsored by The Houston Forum with program support from The Gordon and Mary Cain Foundation.
(3) Kerr, Richard, *Model gets it right - without fudge factors*, Science, May 16, 1997.
(4) Wigley, T. M. L., *The Kyoto Protocol: CO_2, CH_4, and climate implications*, Geophysical Research Letters, 25, 2285-2288, July 1, 1998.

Chapter 3.2 The debates over Climate Sensitivity

• Introduction - History of Climate Analysis: 1824 to 1979

This section is a continuation of the thinking presented in §2.2 on the complexity of temperature equations or models. It is also an introduction into the concept of climate sensitivity of our planet. This is defined as the temperature rise for a doubling of the atmospheric CO_2 concentration. A brief history [1] of climate modeling will be used to open this discussion of climate sensitivity. The review on climate models will range from the very early, one dimensional (1-D) models, with that dimension being altitude, up to the current, gigantic 3-D, general circulation models (GCMs).

In 1824: Joseph Fourier calculated the Earth would be far colder, minus it's natural atmosphere. He was the first to explain that the Earth's atmosphere retains heat radiation. He asked: when sunlight warms the Earth up, why doesn't the planet keep heating up? His answer was that the heated surface emits IR radiation. The atmosphere somehow keeps part of that radiation. He tried to explain this by comparing the Earth to a greenhouse. This was a very major simplification as the main effect of the glass is to keep the heated air from drifting away. Irregardless, this became known as the *greenhouse effect (GHE).*

In 1896, Svante Arrhenius noted that his work on GW required massive calculations of the radiative heat transfer for different levels of CO_2. His model developed temperatures by adding up how much solar energy was received, absorbed, and reflected. However, a true calculation of the so-called GHE would need far more accurate data and a model that included:

· cloudiness;

· the quantities of heat carried from the tropics to the poles by atmospheric and ocean movements

· and updrafts that carry heat from a warmer surface into the upper atmosphere.

Since Arrhenius's "theory" left out some essential factors, the results could not be called predictions. Rather they could give a crude hint of how changes in the amount of gas might affect climate.

Arrhenius's equations went beyond prior work by including *feedbacks*, one being water vapor (H_2O_V). Warmer air would hold more moisture. Since H_2O_V is itself a GHG, the increase of H_2O_V would augment the temperature rise. Arrhenius built into his model an assumption that the amount of water vapor would rise or fall with temperature. He supposed this would happen in such a way that relative humidity would remain constant. That oversimplified the actual changes in H_2O_V, but made it possible for Arrhenius to include this feedback in his calculations.

He came up with numbers, if CO_2 were doubled, of ~ 5 or 6 °C of GW. However, a true calculation of the GHE would need far more accurate and complete data.

In 1928, George Simpson, the first scientist to be referenced in one of Wearts paper: *Basic Radiation Calculations*, noted: "No branch of atmospheric physics is more difficult than that dealing with radiation.

This is not because we do not know the laws of radiation, but because of the difficulty in applying them to gases." Simpson also noted that it was necessary to take into account, in detail, how H_2O_V absorbed or transmitted radiation in different parts of the spectrum. Simpson also calculated how the winds carry the energy from the tropics to the poles, not only as the heat in the air itself, but also as heat energy locked up in the H_2O_V.

In 1938, G. S. Callendar, used 12 layers in his 1-D model, and computed the heat radiation that would come downward from each. As more GHGs were added, the radiation that reached the surface would come from lower, but now warmer layers, while the amount from the cold upper layers was further screened off. This was the true mechanism of the <u>misleadingly named</u> GHE.

In the 1940s S. Chandrasekhar, a great astrophysicist, and others, studied [2] an analogous problem on how energy moved through the "interiors and atmospheres" of stars.

In 1962, one scientist noted that the readers may *blow their minds* on the size of the calculation burden required to calculate the energy budget for a column of air, but fortunately machines are at hand. Digital computers were starting to be applied. Some groups were probing their use to compute the entire 3-D general circulation of the atmosphere. Surely 1-D models would be the foundation on which any **GCM**s would be built. A 3-D atmosphere would be a great many 1-D columns, exchanging air with each other. However, it would be a long time before computers could handle the millions of calculations involved.

Clouds were always the worst problem. How did the cloud cover change with temperature and humidity. What was the correct albedo value to use? Worse, besides the albedo, you needed to know the amount and distribution of cloudiness around the planet. For many decades there had been only guesses. There was nothing better until 1980 with the emergence of satellite measurements. But that only covered existing clouds, and did not consider how these might change in a warming world.

By the late 1970s scientists were now starting to see that the climate system was so rich in feedbacks that a simple set of equations might not give an adequate answer.

In 1979 an NAS panel, chaired by Jules Chaney, was established to compare 1-D and 3-D models. It was somewhat of a surprise to find that the radiative-convective 1-D models results were only 20 % lower than the best GCMs.

This commission set the range for climate sensitivity—the temperature rise for a doubling of CO_2—at 1.5 to 4.5 °C, and flagged the area of cloud feedbacks as one of the weakest links.

Extensive development effort poured into the GCMs over the following 25 or so years. Over 30 such models were developed around the globe. Thousands of papers followed on the application of such models, their shortcomings and on climate sensitivity. In particular the interested reader may find the analysis by Sherwood Idso in NOTE 4 of interest.

• A 1990 *Debate* between a British scientist and three independent American physicists.

A recent news story in the Houston Chronicle reported on an interview[9] with a Dr. C. Rapley, of the British Antarctic Service. He asked two rhetorical questions. "Do you agree that physics is physics? And "If carbon is increasing how can you really deny there's going to be warming?" Rapley challenged the readers, in the arrogant manner that many Britishers are plagued with: if you really knew how physics works, you would *stop arguing on* GW.

As a means to approach this challenge, a rather simplified analysis of the 1990 IPCC assessment will be noted. This is from the book[10]: *Scientific Perspectives on the Greenhouse Problem.* The authors are:
· Robert Jastrow, Columbia PhD, ultimately formed the Goddard Institute for Space Studies (GISS);
· William Nierenberg (1919 - 2000), a former director of the Scripps Institute of Oceanography;
· Frederick Seitz, (1914 - 2008) Princeton PhD in solid state physics and later president of Rockefeller U.
These authors, each with a PhD in physics, viewed the GW, of 1.5 to 4.5°C for the next century—as cited by the 1990 report, as alarmist. They based their analysis only on <u>observed data</u>, and no computer modeling. Their analysis included six assumptions that are listed in NOTE 5.

Their simple analysis was the basis for their conclusion that the IPCC was far too pessimistic, and represented a major exaggeration of the actual physical situation. Well Dr. Rapley these three physicists really know their physics, and they don't deny there is going to be a warming, but no where as big as the IPCC would like the public to believe. However, the IPCC ignored their critique, as the 2001 IPCC report changed the range to 1.4 – 5.8°C.

• A Hissing Contest between James Hansen and the Idso Family

<u>Prologue</u>. In 2007: James Hansen presented alarmist testimony[12] to the House of Representatives in April. Hansen's testimony was entitled *Dangerous Human-Made Interference with Climate* and was based on a paper of the same title[13].

The summary of his testimony includes the claim that our "climate is remarkably sensitive to global forcings – – –. "Huge natural climate changes, from glacial to interglacial states, have been driven by very weak, very slow forcings – – –." He goes on to further claim that today we are "applying much stronger, much faster forcing" as we add CO_2 to the atmosphere. Sherwood Idso and son Craig did not wait very long to provide a critique[14] of this testimony.

The core concept of Hansen's testimony is that the earth "is close to dangerous climate change, to tipping points of the system with the potential for irreversible deleterious effects." However, this contention is neither a self-evident verity nor a proven fact. It is an opinion

One final comment. Hansen sees climate sensitivity, based on climate history, as 3 to 4 °C/Wm^{-2} of forcing. Idsos believe this result to be much too large, based on eight natural experiments—covering time spans both shorter and longer than the one employed by Hansen—to derive a climate sensitivity that is almost an order of magnitude smaller. This indicates the Idsos still see their 1980 and 1998 work as valid.

<u>The Critique</u>.. In a 25 page critique on their web site—43 pages as a quality paper copy—they identified seven key subjects they took major exception to, and another 15 that deserved comment.

Many details on these 22 items will be found in NOTE 6.

<u>The Epilogue</u>   No attempt is made here to cover all of these 22 areas. A few comments are in order. For example they noted that when Hansen's testimony is compared with what has been revealed by the scientific investigations of a "diverse assemblage of highly competent researchers in a wide variety of academic disciplines, we find that he paints a very different picture of the role of anthropogenic CO_2 emissions in shaping the future fortunes of man and nature alike than what is suggested by that larger body of work."

References and Notes

(1) Weart, Spencer, *The Discovery of Global Warming*, Harvard University Press, Sept. 30, 2004. The writings of a Dr. Spencer Weart have been used extensively in this

summary. While Weart is not a GW skeptic, he is rather incredibly prolific writer. While I have a high regard for his effort I am not totally satisfied with his presentation as several times he seems to be acting as a cheer-leader for the *warmers* case.

(2) S. Chandrasekhar, a great astrophysicist, and others, were concerned on how energy moved through the interiors and "atmospheres" of stars. Chandrasekhar developed a set of highly sophisticated equations and techniques. As an indicator on the complexity of such analysis, it takes fusion energy thousands to possibly millions of years to work from the core to the *surface* of the sun. However, this problem was so subtle and complex that he regarded his work as a mere starting point. Indeed astrophysics was too subtle and complex for most meteorologists. They relied on their own shortcut methods to obtain rough numerical results.

(3-8) See NOTE 4

(9-10) See NOTE 5.

(11-14) See NOTE 6

Chapter 3.3 The GCMs: are they adequate for use in policy development?

• Introduction

I have written[1] on this subject before. A very major portion of the global warming case is based on results from computer models. A stronger understanding of the computer models behind the global warming assessments will provide a better position to understand this controversial issue. Hence the purpose of this chapter will be to present an updated situation review on the computer models used in this field. This will concentrate on the 3-D General Circulation Models (GCMs)..

The buildup of the *techno-econometric* model specifications in §2.2, and the discussion of climate sensitivity in §3.2 has been used as a means to convey the complexity of the task at hand. Consider statistical analysis.

· While Multiple Regression Analysis (MRA) has been used in this field, this use has almost been minuscule.

· And Principle Component Analysis (PCA) has also been used in this field, but more precisely misused.

· There surely will be places where MRA and/or PCA can be used, but it was decided long ago that the overall job could only be tackled by very large simulation models.

In 1990, modeling of the global climate was being carried out intensively by 14 major groups in the U.S. and about the same number in the rest of the world. This number has increased slightly since then. Note that these GCMs were originally designed for research planning and education, not for policy development. Hence the question needs to be asked: are they good enough for this more crucial task?

The GCMs are based on dividing the world up into thousands of cells. The chemical and physical processes in each cell would be simulated, at least to some degree, and both material and energy transfer would be permitted between cells. Finally the overall system would then be subject to some external forcing mechanism, such as incremental radiation retention via an increased concentration of GHGs. As impressive as this may sound, the key question remains: is it good enough? And the answer for many applications is: not yet.

On one hand the nature of the development of the GCMs has been striking, and represents the outstanding creativity in the scientific community today. On the other hand this may be a case where we have the maximum mis-use of a technology, as warned about in §1 on the definition of a Goron, namely "those activists-politicians who utilize complex science to isolate rather than illuminate an issue."

In any event in my initial situation review I expressed concern on six aspects of model structure and logic and five areas of model performance. In this current situation review inputs from Wikipedia have been used here.

· Atmospheric GCMs (AGCMs). The AGCMs have progressed to a grid size, for example for one of the Hadley Climate Centre models, of 2.5 to 3.75 degrees. This represents an increase in variables, for a 19 level model, of roughly 250% (~200,000 for the 5X5 model to ~500,000 for the 2.5X3.75 model).
· Ocean GCMs (OGCMs).
· Coupled GCMs (CGCMs).

This spectrum of computer models are all in use.

What follows now are three sections on GCM situation reviews, followed by testimony by three key GCM skeptics

• Situation Review I: Structure and logic concerns about the GCMs, as expressed in 1997.

(1) <u>Model Stability</u>. Separate atmospheric and hydrospheric models were coupled, called the Coupled GCMs (CGCMs), but the simulation had some instability. The practice has been to adjust the amount of heat and moisture flowing between these spheres via the so-called[2] *fudge factors* which have been large.

(2) <u>Model Sensitivity</u>. The variety of GCMs yield a range of forecasts from 1 to 5 °C when forced with a doubling of CO_2—or an equivalent CO_2 doubling (ECD)[3]—a range far too broad to be acceptable.

(3) <u>Role of Water Vapor - H_2O_v</u>. The GCMs would not predict very much warming due to an increase in CO_2 concentration alone. The models rely on major amplification or feedback factors[4] from the estimated H_2O_v in the atmosphere.

(4) <u>Atmospheric Retention of CO_2</u>. The GCMs tend to exaggerate the CO_2 retained in the atmosphere. These models use a constant retention [5] , typically around 56% vs a variation from 21% to 85%.

(5) <u>Impact of Inclusion of Man-made Aerosols (ASLs) in the Models</u>. Proponents claim use of ASLs essentially solves their problems. Skeptics note that inclusion may actually worsen match in North America and Europe, the two regions with the maximum emission of ASLs.

(6) <u>Grid Spacings</u>. These vary, from 10X10 degrees down to 5X5 degrees in latitude and longitude. The atmosphere would also be divided into as many as 20 layers. Even with the 5X5 grid, one still sees regions from San Francisco to Lake Tahoe to Death Valley and back to Los Angeles. Improved models will need more spatial detail to better simulate the processes involved. For example ASLs are released in a very non uniform manner over the globe.

•Situation Review II: Performance concerns about the GCMs:, as expressed in 1997.

(1) <u>Surface Temperature Changes over past 100 years versus GCM Predictions</u>. For the actual increase in Ground Based Data (GBD), the most recent UN estimate of 0.3 to 0.6 °C can be used. In contrast the GCMs have always predicted much more warming. A recent report[6] cited ~ 1.5 °C since 1900.

(2) <u>Satellite Temperature Changes over past 20 plus years versus GCM predictions</u>. See little if any warming in Satellite Based data (SBD), an 18 year record. [As noted in §3.1: Christy effectively defended the satellite based temperature record over the past ~20 years. Several adjustments have been necessary. Christy was confident that the basic differences in temperature trends remain. These show the SDB with very little, if any, temperature trend versus a positive trend for the SDB.]

(3) <u>Night vs Day Warming</u>. The spread [7] between daily maximum and minimum temperatures is getting smaller. One possible reason is thought to be due to a gradually increased level of clouds, which will reduce energy coming

in during the day and help retain more energy at night. Hence most of the warming, that has occurred, has been at nighttime. Daytime temperatures display little or no warming. One report cited values of 0.84 to 0.28 1C or a ratio of 3/1. In contrast the GCMs have predicted a ratio of 11/10.

(4) <u>Winter vs Summer Warming</u>. As there is much more nighttime during the winter would expect more warming. See a ratio of 4.2/1. I know of no GCMs that predicted such a desirable result.

(5) <u>Arctic Warming</u>. Actual results show little warming. For example three studies, based on an average time span of 72 years showed only 0.1 1C warming. Three other reports, on GCM results, over an average time span of 36 years predicted 2.0 1C warming, that is 20 times the warming, in only half the time.

•Situation Review III: Uncertainties in 1997 versus Uncertainties Today

It was easy to get the conviction, back in 1997, that there is an overwhelming consensus from the scientific community that global warming is here and major action must be taken immediately. Indeed, many proponents are repeatedly making this claim every chance they get. Yes, that sentence was written in 1997. But isn't this the same mantra we have heard over and over again in 2006, 2007 and 2008?

Yet this area was endemic with uncertainties. And these were discussed in my initial essay on the GCMs. These comments were based on a paper[8] where the author provided an estimate and bar chart of eight potential climate change forcings, including the basic greenhouse gases. A new, and very similar graph was published [9] in the most recent IPCC report, now with 12 anthropogenic and one solar potential forcings. Estimate from both reports, for the climate forcing by the basic greenhouse gases, in watts per square meter (W/m^2), are shown below.

Item	Expected Value W/m^2	Range W/m^2
1996 Values for CO2, CH4, N2O, CFCs	2.4	2.1 to 2.8
2006 Values for CO2, CH4, N2O, CFCs	2.6	2.4 to 2.9
2006 Values for all anthropogenic forcings	1.6	0.6 to 2.4

Estimated Total Climate Forcings - W/m^2

In spite of nearly ten years of additional effort, the uncertainty for the basic gases has been reduced only modestly. If one considered all the forcing mechanisms—including a mix of ASL forcings and a fairly narrow and limited solar forcing—the uncertainty would again be rather huge.

There were also major problems with the computer models including major uncertainties in the background processes and on how to simulate these. One report[10] by the noted sceptic, Richard Lindzen, charges the amplification mechanisms used in the GCMs depends on what is likely to be a severe misrepresentation of the relevant physical processes. A second report[11]—by a writer who has been more than friendly to the proponents side in the past—summarized: we should not be surprised on the shortcomings of the GCMs given the number of climate processes that are poorly understood or totally unknown.

So much for my initial situation review. The key question have these concerns and others been confirmed and satisfactorily worked out over the past ten years? Read on.

As far as the quality and validity of the modeling field today, key inputs from selected skeptics, will be used. This will start with inputs from Richard Lindzen— to follow up on his concerns expressed on amplification processes— then followed by inputs from Sherwood Idso, and inputs from a European scientist, a Dr. Hendrik Tennekes.

• Key Skeptic I: Inputs from Dr. Richard Lindzen

Lindzen is an atmospheric physicist and the Alfred P. Sloan Professor of Meteorology at MIT. He is, perhaps, the leading academician in the GW debate, and is a member of the National Academy of Sciences. He is a recipient of the AMS's Meisinger, and Charney Awards, and AGU's Macelwane Medal. He is a member of the National Research Council Board on Atmospheric Sciences and Climate.. Yet he has been attacked as a shill of the oil industry and incapable of having his own views. He has over 229 publications on such subjects as Hadley circulation, monsoons, planetary atmospheres, hydrodynamic instability, mid-latitude weather, global heat transport, the water cycle and ice ages.

Lindzen is skeptical on the GW issue in general, and the GCMs in particular. Several of his papers are reported on in NOTE 7.

A fitting conclusion to these remarks by Lindzen is provided in a rather broad 2006 editorial[18] by Lindzen, in the WSJ, titled *Climate of Fear.*

The secondary headline noted the GW alarmists intimidate dissenting scientists into silence. This editorial covers:.

(a) Lindzen asks "how can a barely discernable, one-degree [°F] increase since the late 19th century possibly gain acceptance as the source of recent weather catastrophes?" His answer is that "ambiguous scientific statements about climate are hyped by those with a vested interest in alarm – –." He asks who puts money into science where there is nothing really alarming? He notes that "scientists who dissent from the alarmism have seen their grant funds disappear, and their work derided and themselves libeled as industry stooges."

(b) He noted how the process—of new paper, letters by critics, and letters in response by the original author all in the same journal—was changed. He noted several hastily prepared papers appeared, claiming errors in our study, with our response delayed months or longer, allowing it to be noted as "discredited.".

(c) He also noted that alarm, rather than genuine scientific curiosity, "is essential to maintaining funding. And only the most senior scientists today can stand up against this alarmist gale, and defy the iron triangle of climate scientists, advocates and policy makers."

•Key Skeptic II: Inputs from Dr. Sherwood Idso (SI).

Idso was reported on extensively in §3.2, on climate sensitivity. He has been equally active on the subject of modeling. His web site[19] is strongly recommended. It provides a highly useful—on a weekly basis—set of situation reviews, editorials and journal reviews, on a wide array of issues on climate change. This site includes a detailed Subject Index. A few examples follow.

(1) Model Inadequacies - Radiation. In a major situation review[19a] Idso argues there are major inadequacies on the handling of the Earth's radiative energy balance, plus numerous other telling inadequacies from the exclusion of pertinent phenomena. As such "there is no rational basis for any of the IPCC inspired predictions of catastrophic climatic changes due to the continued anthropogenic CO_2 emissions."

(2) A UK Modeler's Perspective. In this technical paper review[19b] Idso reports this writer lists many unresolved features such as "ocean eddies, gravity waves, atmospheric convection, clouds and small scale turbulence." This writer went on to state that "the full spectrum of spatial and temporal scales exhibited by the climate system will not be resolvable by models for decades, if ever." Idso comments that "it is strange indeed that the

present-day outputs of these vastly imperfect tools are considered by some to be so sound as to justify a complete restructuring of the way the world produces and uses energy." He asks, perhaps rhetorically: "have such folks all gone mad???"

(3) <u>Simulating the Past</u>. In this technical paper review[19c] Idso reports the writer reviews the results of 19 CGCMs attempt to simulate the Sahel drought signal[20]. Idso observed that all 19 models "failed to adequately simulate the basic characteristics of 'one of the most pronounced signals of climate change' of the past century (the Sahel drought of the 1970s-90s." This failure of the "ideal test" for evaluating the models "would almost mandate that it would not be wise to rely on their output as a guide to the future."

(4) <u>Climate Model Inadequacies - Clouds</u>. In a major situation review[19d] Idso notes that there are a multitude of problems that restricts our ability to properly model cloud related processes. This, in turn, "restricts our ability to simulate future climate with any degree of confidence in the accuracy of the results.".Hence. "The model inspired specter of catastrophic CO_2 induced global warming that looms on the horizon is but a paper tiger, totally clawless and devoid of teeth."

• Key Skeptic III: Inputs from Dr. Hendrik Tennekes (HT)

Tennekes is the former director of research at the Royal Dutch Meteorological Institute and currently a professor of aeronautical engineering at Penn State. He has written two books on aeronautics. In 1972: A First Course in Turbulence, and in 1997: The Simple Science of Flight, both published by the MIT Press. The subject of turbulence—a field of importance in fluid mechanics and boundary layer considerations—would be a field of major utility in understanding GCMs. While a strong proponent of scientific modeling, he is an equally strong opponent of climate modeling and the GCMs. HT was forced out of his Dutch post due to his very strong comments on climate science in general, and the GCMs in particular.

(1) <u>Early Views on the GCMs</u>. I first came upon HT's work in an essay[21] posted on the Roger Pielke Sr. Web site. Pielke is a scientist that I have come to respect and admire and as such I periodically peruse his site. I then looked at Wikipedia[22]. This provided references to his two books, and four external links. Some highlights for each link follow.

(a) *A skeptical view of climate models* - In this essay[22a] HT noted he protested he had "become very sensitive to everything that smells like an orthodox belief system." He added "the moment one accepts a dogma, one stops being an independent scientist."

(b) *The Lorenz paradigm and the limitations of climate models* - The above essay noted that HT had been involved in a problem for 30 years, namely the problem of a finite prediction of complex deterministic systems. This is the very first problem defined by Edward Lorenz[22b]. In 1986 HT gave a speech entitled: *No Forecast is Complete Without a Forecast of Forecast Skill.* This concept is still not properly accounted for by the majority of climate scientists. HT closed this essay with the note "that the task of finding all nonlinear feedback mechanisms the microstructure of the radiation balance probably is at least as daunting as the task of finding the proverbial needle in the haystack." He went on. "The blind adherence to the harebrained idea that climate models can generate 'realistic' simulations of climate is the principal reason why I remain a climate skeptic. One final input. "From my background in turbulence I look forward with grim anticipation to the day that climate models will run with a horizontal resolution of less than a kilometer. "The horrible predictibility problems of turbulent flows will then descend on climate science with a vengeance."

(c) *The limits of predictability* - Here Wikipedia refers to an editorial and a chapter in a book[22c] by a Lawrence Solomon, that Hendrik Tennekes, more than any other critic, has challenged the GCMs that climate scientists have, and are still constructing. He argues what is needed is a different approach to this science, an approach that recognizes inherent limits in such scientific tools.

(d) *A Personal Call for Modesty, Integrity and Balance* - This essay[22d] by HT, although posted in 2007, notes that HT's plea goes back 17 years..

(2) <u>Current Views on the GCMs</u>. In the past HT's feelings were one of concern.Today his concerns are more on anger, primarily focused on the IPCC. This falls in two areas:

(a) their CO_2 fixation and their pre-occupation with CO_2 emissions.

(b) the monopoly position that GCMs have achieved in climate research.

He sees this as strategy by the IPCC, not science. He notes there are many other areas demanding more research, but not necessarily by more, or bigger GCMs. He notes that GCMs have been running for 20 years now, but the multiplicity of models can't be made to agree on anything except <u>a possible relation between GHGs and a slight increase in globally averaged temperature</u>, and <u>a likely link to fossil fuels use</u>. But that is the end of the consensus.

HT notes one example, out of many, of a major short-coming: the GCMs do not include feedbacks between changing farming and forest practices and the atmospheric circulation. For this and other reasons they can't agree on precipitation patterns. But <u>precipitation is far more relevant to global food production than a slight increase in temperature.</u>

(3) <u>Air circulations</u>. As a related item, HT has commented on such phenomena as the jet stream, the Polar Vortex and the Arctic Oscillation (AO). He has quoted[22d] a Dr. John Wallace—a Professor of Atmospheric Sciences at the University of Washington who has been Co-Director of the Program on the Environment at this university, and has been a Member of the Committee on the Science of Climate Change for the National Research Council/National Academy of Sciences—that "there is not a beginning of consensus on a theory of the AO." While the AO may sound rather quaint and unimportant, it is one of several oscillations in climate that are indeed important. The El Ni_o-La Ni_o-Sothern Oscillation is surely the best known of these phenomena. Wallace noted[23] that the AO affects sea-ice on many different time scales.

HT observed: without "an established relationship between rising GHG concentrations and systematic changes in the AO, one cannot possibly make inferences about changes in precipitation patterns." As a result, HT went on, we do not know, and for the time being cannot know anything about changing patterns of clouds, storms and rain.

Idso, in a recent situation review[24], amplified the situation with regard to precipitation. He studied 14 papers relating to various areas of precipitation. Each study relied on inputs from a variety of GCMs. All these models had extensive shortcomings. To cite one example, consider the models inability "to correctly simulate one of the largest and most regionally important of Earth's atmospheric phenomena - the tropical Indian monsoon. "After more than 70 years of trying to remake the models into better predictive tools, one would surely have expected some improvement in this regard, even if only by accident. "That there has been absolutely none is a sad commentary indeed on the state of the climate modeling enterprise."

Finally HT notes that modeling is the basis of forecasts of climate change, but he argues this modeling has little utility. He states: "There exists no sound theoretical framework for climate predictability studies." HT added : "We only understand 10% of the climate issue."

• **Conclusions**.

While all of the above is not proof that the huge GCMs—used so extensively to make the warmers case—are seriously flawed, it is surely food for thought on these tools.

· The initial concerns, expressed in this writer's situation review of 1997 on model structure and performance, are still very much in play, even though the models have been changed substantially.

· The views of three key skeptics have been noted directly and several others indirectly. Their lifetime publications, speeches and comments give the nature of these scientists views on the quality of the GCMs.

· Although a multiplicity of GCMs have been running for 20 years, one scientist argues they agree on very little except <u>a possible relation between GHGs and a slight increase in globally averaged temperature</u>, and <u>a likely link to fossil fuels use</u>.

· One example of a major short-coming: the GCMs do not include feedbacks between changing farming and forest practices and the atmospheric circulation. As such they can't agree on precipitation patterns. But <u>precipitation is far more relevant to global food production than a slight increase in temperature.</u>

· The inputs in this chapter are surely not proof that the GW issue is heading down the wrong highway due to its reliance on the GCMs, but it again is *food for thought*. It suggests it may be time to stop and get off this speedway for awhile, to look for a map, and to double check our directions.

· The claim that we face an imminent catastrophe is unfounded and inappropriate. The extensive use of alarmism in general, by the supporters of the warming position, do their case a major disservice..

· In the title to this chapter we asked: Our the GCMs adequate for use in policy development? Our answer would be No, they're surely are not!

References and Notes

(1) Westbrook, Gerald T., *Global warming Models: Are they Adequate for Policy Development?*, IAEE Newsletter, Summer, 1997.

(2) Kerr, R., *Climate Modeling's Fudge Factor*, Science, **265**, 9-9-94

(3) Each greenhouse gas contributes a unique amount to the overall greenhouse effect. As such the impact of a doubling of $[CO_2]$ can be defined by CO_2 alone, or by the sum of the contributions from all of the gases. This is referred to as the ECD --- the Equivalent CO_2 Doubling.

(4) Lindzen, R., *Errors Hurt Global Warming Theories*, NY Times, 11-30-90

(5) Keeling, C. D. et al, *Atmospheric Retention of* CO_2, Nature, **375**, 6-22-95

(6) Mitchell, J.F.B., et al, *On Surface Temperature, Greenhouse Gases, and Aerosols: Models and Observations*, Journal of Climate, **8**, 10-95

(7) Karl, T.R., et al, *Asymmetric Trends of Daily Maximum and Minimum Temperature*, Bulletin of the American Meteorological Society, **74**, 1993

(8) Schwartz, S., et al, *Uncertainties in Climate Change Caused by Aerosols*, Science, **272**, 5-24-96

(9) *A Report of Working Group I of the Intergovernmental Panel on Climate Change, Summary for Policymakers*, 2007, Figure SPM.2..

(10) Lindzen, R., *Absence of a Scientific Basis*, National Geographic Research & Exploration, **9(2)**, 1993

(11) Kerr, R., *Dark Clouds Promise Brighter GCM Future*, Science, **267**, 1-27-95

(12 - 17) See NOTE 7.

(18) Lindzen, R., *Climate of Fear*, The Wall Street Journal, April 12, 2006.

(19) See www.co2science.org,

(19a) *Climate Model Inadequacies (Radiation) - Summary*, Last updated January 18, 2006

(19b) *A UK Modeler's Personal Perspective on the Status of Climate Modeling*. A review of the technical paper by Williams, P. D., *Modeling climate change: the role of unresolved processes*, Philosophical Transactions of the Royal Society, A, **363**: 2931-2946, 2005. Reviewed May 17, 2006.

(19c) *Simulating the Past: A Test of State-of-the Art Climate Models*. A review of the technical paper by Lau, K. M., et al, *A multimodel study of the twentieth century simulations of Sahel drought from the 1970s to the 1990s*, Journal of Geophysical Research **111**: 10. 1029/2005JD006281, 2006. Reviewed August 30, 2006.

(19d) *Climate Model Inadequacies (Clouds) - Summary*, Last updated January 25, 2006

(20) The Sahel drought, from the late 1960s to early 1980s, created a famine that killed a million people and impacted many millions. The economies, agriculture, livestock and human populations of many parts of northwest Africa during this drought were severely impacted.
Because the Sahel's rainfall, the West African Monsoon, is heavily concentrated in a very small period of the year, the region has been prone to droughts since 3000BC. Although this record is far from complete specific droughts or dry periods have been noted at 1640, 1680s, 1790 - 1870, 1910s, 1940s and finally the 1960s to the early 1980s.

(21) See the Roger Pielke Sr. Web site at: http://climatesci.colorado.edu. Select March 2007 entry and scroll down to find: *Some Fresh Air in The Climate Debate* - An Op Ed by Hendrick Tennekes. (22) See Wikipedia web site: www.wikipedia.org/wiki/Hendrick_Tennekes. This was last updated February 21, 2008.

(22a) This link led to an undated paper by Tennekes, published on the SEPP site about ten years earlier.

(22b) In this essay Tennekes also reviews the thinking of Karl Popper

(22c) Lawrence Solomon is a writer for the National Post/Financial Post in Toronto. In addition to being a columnist, Solomon is also a world renowned environmentalist and conservationist. He is the founder of Energy Probe Research Foundation and executive director of its Urban Renaissance Institute and Consumer Policy Institute divisions. Recently he has completed a series of editorials which is now out in book form entitled - *The Deniers: The World Renowned Scientists Who Stood Up Against Global Warming Hysteria, Political Persecution and Fraud** And those who are too fearful to do so,* Richard Vigilant Books, February 28, 2008. Two chapters will be noted here.

(i) The limits of predictability -- The Deniers Part VIII.(ii) Look to Mars for the truth on global warming -- The Deniers Part IX.

Solomon wrote, in Chapter IX,, that the climate of Mars is the warmest it has been in decades, even centuries. He quoted a NASA scientist, William Fieldman, that Mars could be just coming out of an ice age. "With each passing year more and more evidence arises of the dramatic changes occurring on the only planet – – – apart from Earth, to give up its climate secrets."

Solomon also quotes a Dr. Habibullo Abdussamatov (HA)—head of St. Petersburg's Pulkova Astronomical Observatory space research laboratory—"Mars has global warming, but without a greenhouse and without the participation of Martians." He went on: "These parallel global warmings—observed simultaneously on Mars and Earth—can only be a straight-line consequence of the effect of the one same factor: a long term change in solar irradiance." HA goes further debunking the very notion of a greenhouse effect. (22d) See the Roger Pielke Sr. Web site at: http://climatesci. colorado.edu. Select January 2007 entry for *A Personal Call For More Modesty, Integrity and Balance* - by Hendrik Tennekes.

(23) Rigor, Ignatius G., Wallace, John M., *Variations in the age of Arctic sea-ice and summer sea-ice extent,* Geophysical Research Letters, **V31**, May 8, 2004.

(24) Idso, Sherwood, *Weather Extremes (Precipitation - Model Inadequacies) - Summary,* Last updated May 4, 2008. See www.co2science.org.

4. All About Florida

In my first book, this section was limited to the Florida election system, voter eligibility, ballot counting, ballot inspection and other related subjects. A very creative limerick[1] helped define the importance of this election. Ditto for the ballot and the chad. It served to set the proper tone for this subject. This chapter also had noted that it would be appropriate for the Epcot Center to open a technology center on "Chads" at their campus in Orlando. The Duplantier limerick could be prominently displayed at this center.

Recount Dracula
by F. Duplantier

There once was a man named Vlad
Who was known for a habit he had:
With some pride in his nailing,
When he took to impaling
No one ever would challenge his chad.

In addition to the chad, as an item to be preserved for future generations to see, I had argued there is something even more valuable to preserve — and that was the name Goron. At that time it was not clear if the HG would be around for any future political activity. Many reports had declared his political career over. As such, there had been some suggestions that *the Gorons rename their organization*. This should be discouraged. After all the name Goron is such a perfect fit for all liberals, environmentalists and the global warming crowd. And it rhymes with the other alien group the Morons. So this name should be preserved. It is short, crisp, descriptive, memory provoking and too perfect a fit to abandon. It should be used over and over as a perfect, yet simple, synonym for unlimited government, robotic policy wonks and politicians that embrace endless regulations, with a special extra regard given to international regulations

Again, in my first book, the *Florida battle* was highlighted, as a means to remind the readers of how determined, tenacious and even ruthless the Gorons had been, and will continue to be. I noted that, while this band of aliens lost this battle in Florida, count on them to carp about the *injustice* of this election for many years. They have not disappointed me on that count. While this 2000 election may have settled the battle with the Gorons in the short term, the Gorons have not gone away.

Many individuals moved into this Florida battle. Two deserve comment here:

· Warren Christopher, former secretary of State, has been described by the comedian Al Franken as "half beyond reproach and half dead." And now Franken is a senator from Minnesota, having clearly learned from the Florida battlefield.

· Joseph Lieberman — this gentleman may not be half dead, but his voice surely must be. In any event I no longer see him as a Goron.

A brief update on the Florida elections, systems and people will be covered here in § 4.1. Now there are other issues of import in Florida beyond their election systems. Two additional subjects, of particular importance to Florida, will be added to this new book namely: § 4.2 on hurricanes and § 4.3 on sea level rise.

Chapter 4.1 The Florida Invasion — an Update.

• Florida election of 2000.

When the election was over on November 7th, 2000 the victory in Florida went to Gore, then to Bush, then to On-hold.

Now the Florida Supreme Court vote was 4 to 3 for a recount. For good liberals, like four on the Florida court, when procedural justice is inadequate something new must be invented. And these Gorons, being from outer space, came up with *cosmic justice* which allows constant refinement, and "ordered another recount — selective, standardless, seat-of-the-pants[2]"

In turn, the U.S. Supreme Court ruled, by a vote of 5 to 4, that the Florida recount plan was so arbitrary and capricious as to be not only unfair, but also unconstitutional. While the case was remanded back to the Florida court for further proceedings, at the same time the five U. S. justices said that legal deadlines would make a constitutionally sound recount impossible. In addition, three of the justices—William Rehnquist, Antonin Scalia and Clarence Thomas— concluded the Florida Supreme Court had violated both the U. S. Constitution and federal law in ordering a recount.

Finally, on November 26th, 2000, the Florida Secretary of State, a Katherine Harris, announced [3]: "In accordance with the laws of the

State of Florida, I hereby declare Governor George W. Bush the winner of Florida's 25 electoral votes for the President of the United States." Bush received 537 more votes than Vice President Albert Gore.

• Katherine Harris - U.S. Congresswoman

The Gorons never forgave Harris for not calling for a recount, As such, they attacked her rather ruthlessly and she became the most talked about election official in the state. The attacks included [4] an HBO movie *Recount*, that mocked her without mercy.

In spite of these attacks she had a fairly successful political career with election as a U.S. Congresswoman in 2002 and 2004. During this latter campaign a local Sarasota resident [5] "swerved off the road and onto the sidewalk, directing [his auto] at Harris and her party." No one was injured, but this driver was arrested and charged with "assault with a deadly weapon."

During her second term several scandals emerged that—rightly or wrongly—were coupled to her office. These events surely contributed to her defeat for U.S. Senator in 2006.

• Miriam M. Oliphant (MMO) - Broward County elections supervisor

Now during this period another Florida politician emerged, from another voting debacle. However, this time the "most talked about politician in Florida" was a Democrat. Her odyssey has been summarized [6], in part.

· January 2001, MMO is elected as elections supervisor.

· September 12, 2002, a new $32 million voting system, that replaced the one blamed for the debacle in 2000 presidential election fails first test in statewide gubernatorial primary - - - with massive failures in Miami-Dade and Broward Counties - - -."

· January 25, 2003, is under the microscope after overseeing two troubled elections - - - and reviving memories of the 2000 election fiasco.

· April 24, 2003, MMO announces she will seek a second term.

· October 26, 2003, the secretary of states office finds her office is poorly managed and unprepared to handle coming election. Gives MMO until 11/10/09 to come up with plan to correct problems.

· November 21, 2003, Governor Jeb Bush suspends MMO.

· July 17, 2004, MMO files to run for her old job. State Senate continues hearing on whether she should be reinstated.

· May 4, 2005, Senate accepts report upholding Governor Jeb Bush's suspension of MMO.

This turmoil[7] has left municipal and county official with little confidence in MMO's "ability ever to produce chaos-free elections in Broward County, home of lost ballots, missing voting machines and poll workers who close precincts early when they are too tired."

• Where oh Where Have Our Election Systems Gone?

Now there is little point in trying to decide which of these two politicians were talked about the most. Rather the question is if they, and the voting systems they have been involved in, are harbingers of the future, on a national scale. Indeed, after the 2008 election, with key developments in Illinois and Minnesota, to name only two states, the question is most appropriate?

A recent news item announced [8] that a recently formed group, that was set up to investigate and document voting fraud in the 2004 election, was under attack. This group, The American Center for Voting Rights (ACVR) has as its objective "to make it easier for people to vote, and harder to cheat." It has already looked into election fraud in Ohio, a battleground state in the 2004 presidential election. However, it is being attacked due to many connections to the GOP.

In the Minnesota election another secretary of state, a Mark Ritchie, was very key in getting Franklin elected. Richie has been flagged [9] as being backed by far left billionaire George Soros and groups like Acorn.

Bernard Goldberg, the former CBS insider, has observed [10] that we no longer live in the United States of America, "we live in the United States of Entertainment."

Another blog - *The ecologic Powerhouse* has had some critical observations on our election process. First of all the writer observes[11] we no longer have an "American" electorate. Half the eligible voters don't vote and "the other half has a dirty little laundry list of special interest litmus tests - - -." Next he lists the attributes of a desirable leader:

· one that was an outstanding American;
· one with good leadership skills;
· one with strong family values;
· one with a sense of decency;
· one with firm moral values; and the
· candidate with America's best interests at heart.

However, today, a candidate must be more. Now he or she must be the "best dancing chicken."

He went on: "the fact is we aren't looking for a good leader. "We are looking for a good follower, - - - a good promise maker, not a good promise keeper." Please note this was written in 2007, well before Barack Obama became president. Further "we are not looking for anyone who will tell us the truth. "We want to be lied to." *The ecologic Powerhouse* writer argues that "Americans have become their own worst enemy. "They alone will destroy America – and they are well on their way." Later: "we are bankrupting our nation with ever increasing government waste, corruption and special interest pandering - - -." Again this conclusion was written in 2007, well before the tidal wave of spending by the new administration.

• Conclusions.

In closing it is hard not to be cynical about our election process. Just as we are living in a tidal wave of propaganda, we are also living in a tidal wave of operators, of people who have gamed the system or who will game the system. And as with propaganda we will need very many floodlights focused on this situation to bring visibility to the public.

References and Notes

(1) Duplantier, F., *Recount Dracula*, Eco-Logic On-Line, December 15 2000.
(2) Krauthammer C., *Gore's campaign lived, died by courts*, Houston Chronicle, Dec. 15 2000.
(3) *Katherine Harris*, as posted on NNDB tracking the entire world. See: http://www.nndb.com/people/067/000038950/.
(4) Hannity, Sean, *Katherine Harris Unhappy with 'Recount'*, FoxNews. See: http://www.foxnews.com/story/0,2933,362303,00.html.
(5) *Katherine Harris*, from Wikipedia. See: http://en.wikipedia.org/wiki/Katherine_Harris.
(6) *Times Topics - Miriam Oliphant*, The New York Times, October 16, 2009
(7) Canedy, Dana, *Florida Official Criticized Amid Familiar Election Chaos*, The New York Times, January 25, 2003.
(8) Rhodes, Kathleen, *Liberal Bloggers Pounce on Voting Fraud Watchdog Group*, CNSNEWS, March 31, 2005.
(9) Friedman, Brad, *RNLA Issue Fact-Free Letter Claiming Desperate Franken is Stealing U.S. Senate Seat*, February 15, 2009. See http://bradblog.com/?p=6918. This blog contains a copy of the letter from RNLA (Republican National Lawyers Association) by Michael Thielen. Amongst other inputs in this letter RNLA is defined as the leading Republican group fighting vote fraud in the USA.

(10) O'Reilly, *Bill, Polls Reveal Viewer Opinions on Media*, an interview with Bernard Goldberg, The O'Reilly Factor, as reported on LexisNexis News home Page, November 2, 2009.
(11) Williams, J. B., *The American Election Process: Reduced to Dancing Chickens*, The ecologic Powerhouse, August 20, 2007.

Chapter 4.2 In Hartford, Hereford and Hampshire, Hurricanes Hardly ever Happen.

What Does a Dust Bowl Refuge Know About Hurricanes?

Before discussing the growth of my interest in, and overall understanding on hurricanes (HCs) and tropical storms (TSs)—and the various skills of Professor Henry Higgins—it is appropriate to define the types of storms of interest. An ocean storm becomes a hurricane when the rotational speed exceeds 73 miles/hour, sustained[0] for over ten minutes. Tropical storms start when speeds exceed 39 miles/hour.

Category or Type	Miles/hour
Named Storms	> 39
Tropical Storms	39-73
Hurricanes	> 73
Category 1	74 – 95
Category 2	96 – 110
Category 3	111 – 130
Category 4	131 – 155
Category 5	> 155

Definitions for Atlantic Storms
The Saffir-Simpson Hurricane Wind Scale

• **Initial Awareness**

Now many of you may not realize that Henry Higgins[1.1], a linguistic expert, was also an expert on hurricanes. At least as expert as some—such as Robert Kennedy Jr.— who claim to be an expert, not only on HCs, on energy, on the hideous nature of coal, and also on global warming (GW) and the relationship of HCs to GW. Note that Higgins was absolutely correct on his claim about *H, H & H*, with only one HC in England since 1703, and even that Great Storm[1.2] of October 15th, 1987 may actually have been a TS, not a HC. However, gusts up to 122 miles per hour caused havoc and major damage across London, 18 people were killed and 15 million trees were blown down.

Like so many Limys, Higgins spent some time out in the *provinces*. Would you believe the University of Saskatchewan? Would you also believe I was his prize HC student?

When I grew up in Saskatchewan there was no TV. This meant, among other things, no CNN and no Weather Channel. And hurricanes were simply not big news out there during the dust bowl, the depression and World War II. My first HC input may have been from the Readers Digest, in the 1940s, on the Long Island Storm. This was a Category-3 HC, that hit in 1938. It was hard for this dust bowl refuge to believe water levels so high—as to swamp ones car—could occur. They did, and indeed, much higher than that. More details on this storm follows, which killed over 700.

• Growing Awareness

I moved to Ontario in 1955. I can recall a severe thunderstorm in June of 1957. I didn't know then that this was the tail end of HC **Audry.** This was the strongest June HC on record. It traveled up from the Gulf of Mexico, across Pennsylvania and into Ontario.

By the 1960s I had moved to a large petrochemical company, with head offices in central Michigan. In the 60s, 70s and early 80s I took very many trips—to the Houston area, to the Baton Rouge/New Orleans area, and a few to Florida and Puerto Rico—and encountered many torrential downpours. I first got acquainted with the power of a tropical storm, when—after an all day meeting without any radio or TV inputs—I and my colleagues emerged from our Houston offices, near the Galleria, right into what turned out to be tropical storm **Claudette.** At that time I was unfamiliar with the really serious situations that can be caused by *mere tropical storms.* So we drove around, rather naively, for about an hour, in six inches or more of water, looking for the restaurant where we had planned to continue our meeting. Never found it. Finally came to the conclusion we had better head for our hotel. Again, never found it. Finally we looked for any hotel and were very fortunate to find one near the Astrodome. About 25 miles from the Astrodome, Alvin Texas later reported [2] to have received 42 inches of rain from this storm.

However, it wasn't until tropical storm **Allison** occurred in 2001, that I fully appreciated the danger and trauma of a *mere tropical storm.* Allison caused many deaths, unbelievable flooding and huge property loss. The greater Houston area received about 35 inches of rain over 4 - 5 days from Allison as it *refused* to go away. The conventional steering currents were not operative, so the storm staggered north, perhaps a hundred miles, then did a complete u-turn back to hit the area again. Houston driving was essentially knocked out. Impacts included [3]:

· 18-wheelers floating in every direction possible;

· damages approaching 2B$ to the many hospitals and research facilities at the Texas Medical Center, including losses of thousands of research animals trapped in the basements; and

· a terrifying drowning of a women trapped in an elevator as she attempted to get to her car in an underground garage.

This tropical storm surely contributed to my education. Not only did it point out that a storm doesn't have to be a HC to be of major danger, but also that steering currents were perhaps the most important component of any forecasting output.

I was transferred to Houston in January 1982, and experienced my first hurricane, **Alicia**, in 1983. Even though my home is 75 miles from the Gulf, this was as close—six miles from its eye as it moved across Houston—that I ever want to get. However, severe "freshwater flooding was minimized by Alicia's fast movement inland[4]."

TV watchers may remember pictures of downtown Houston where "shards of glass became deadly missiles." It was literally raining glass. What happened was the storm would pick up loose gravel from rooftops and hurl the stones, like machine gun bullets, into rooms at lower levels. Thousands of windows were broken in this manner. The building codes have since been changed to prohibit loose gravel on rooftops.

I started to study hurricanes, in depth, in 1998 when I presented the first of four papers on global warming at the Offshore Technology Conference (OTC) [5.1-5.4] in Houston. Scores of documents were studied, including the book *Isaac's Storm*, by Erik Larson [6]. Isaac Cline was the U. S. Weather Service Chief at Galveston in 1900. This book gives details on the Galveston storm that hit that island on September 8 and 9, 1900, and killed over 6000. Details on this storm follow.

Next, we have had **Katrina** and then **Rita**, my second hurricane. Indeed, my family and I were evacuees from Rita, for all of eight hours. This led to another important addition to my HC education: evacuation from a TS/HC can be extremely difficult if not impossible. We got 35 - 40 miles north of Houston in this eight hour time span. That is five miles per hour. I finally convinced my caravan— two vehicles, myself and my wife, two adult offspring and five dogs—to turn around. Reasons: Rita had edged eastward a bit and now looked like it would come ashore on the TX/LA border. Also we could turnaround now, but in the very near future the state would make all lanes of the expressway point north; and if we didn't turnaround now, we would have to spend the night on the highway.

• Pseudo Hurricane Experts

Since Katrina and Rita, we have had many experts from the entertainment field pontificating on the couple between hurricanes and global warming. Perhaps the justification for this testimony was best given by the cartoon[7] ***Shoe* by Jeff MacNelly:**

· Ace reporter *for the Tattler Tribune, Perfesser Cosmo Fishhawk* says: "But Senator..."

· And the senator replies: "Look! I don't pretend to know all the answers about global warming, government propaganda and immigration reform- - that's what we have movie stars for!"

Sean Penn, Leonardo DiCaprio, Barbra Streisand and their like have not disappointed. Nor have the politicians such as Robert Kennedy Jr., and the HG. All of these Gorons continue to pontificate that such hurricanes were caused by global warming. Now one might ask what does this set of refuges, from the planets Gore or More, know about hurricanes? Clearly, this dust bowl refuge, need not be concerned that his credentials were lacking. I would argue that I have done far more research on these storms than these Gorons. Further, I would note, that I am not presenting myself as an expert on this subject, but rather as a reporter. And finally, I would humbly argue that I can report the news and the science of such storms more completely and more objectively than the above Gorons including their leader, Albert Gore Jr..

Key Statistics

Let us look at some statistics on hurricanes. First of all the year 2005 was awesome. Not only did we have Katrina and Rita, but the Atlantic Basin saw 28 named storms as noted below

Storm Type	Location	Count
Named Storms	Atlantic Basin	28
Tropical Storms		13
Hurricanes		15
Category 1 – 2		8
Category 3 – 5		7
Hurricane total	Direct U. S. Hits	6

The 2005 Atlantic Storm Season

A total of 15 of these storms were hurricanes, with six representing direct hits on the U. S. As noted below, these six hits in 2005, were well above the 1851 to 2004 average of 1.77 hits. That was the average from data spanning 154 years. However the record is seven hits that occurred in 1886. Six hits also occurred in 1916, 1985 and 2004.

Location	Direct Hits	Average Hits/Year
United States	273	1.77
Florida	110	0.71
Texas	59	0.38
Louisiana	49	0.32

Hurricane Hits from 1851 to 2004

A comparison of HC hits per decade is also of interest. The records show the USA would see 17 to 18 on the average. Florida and Texas/Louisiana combined are essentially the same at seven hits per decade, with Texas getting four of these and Louisiana three. For the New Orleans area—counting any storm within 85 miles, in the last 120 years— there have been 12 major storms, an average of one per decade. So the odds for New Orleans are not that bad, but as the saying goes, it only takes one.

There is also a cyclical pattern for severe Atlantic HCs over the period 1946 to 2005. For example—using a three average to help isolate this pattern—one sees about 2½ periods: 1946 - 70; 1970 - 1991 and 1991 to 2005.

However, there is not much trend shown. For example—for Florida hits, using a five year average to help reduce the data load, over the period 1886 to 2005—one sees essentially a random pattern.

Key Hurricanes

• The Galveston Hurricane
<u>Prologue to the Galveston Hurricane</u>. In June of 1982 we headed down to Galveston. This trip was the first chance we had to explore the island. And it was the first time I began to hear about the Galveston Hurricane. And it was the first time I heard about Larsons book[6.1] *Isaac's Storm*.
Larson had cited two earlier killer hurricanes that hit 150 miles south and west of Galveston on Matagorda Bay.
· The September 16, 1875 storm shoved an immense dome of water through Indianola, pushing the waters of this bay inland for 20 miles. This Storm took 176 lives.
· The August 20, 1886 storm completed the destruction of Indianola. So many people were killed that the town was abandoned.

Additional extreme weather events are noted in this book and reported on in NOTE 8.

In 1891, Cline was asked about Galveston's vulnerability to extreme weather. He wrote: "The opinion held by some that Galveston at some time will be seriously damaged by some such disturbance is simply an absurd delusion." He made this statement in spite of the fact that the elevation of Galveston Island, was an almost trivial 8.7 feet above sea level.. And he made this statement in spite of the two killer hurricanes that destroyed Indianola. He had publically stated that such HC s would be steered north before hand. In any event this was his *model* of these gulf coast storms. However, his model was fatally flawed, or politically influenced, and it resulted in 6,000 to 10,000 deaths.

Larson suggests there was a "scent of boosterism" behind Cline's 1891 paper[6.2], and that he was writing an article that Galveston promoters would be happy to see. Could this have been a harbinger of things to come? Could Cline have been the first government weather scientist to prostitute himself by writing favorable papers on the weather for his sponsors?

In any event this storm became organized west of the Cape Verdes Islands around August.27 and moved due west. By September 3, Jamaica, to the south, was hit by torrential rains, as was Santiago Cuba to the north. Cuban authorities saw the approach of this storm and believed it would cross Cuba then gain strength, but head out into the Atlantic. It crossed Cuba on September 5, then cut into the Gulf. However, forecasters—at the U. S. Weather Bureau in Washington—telegraphed New Orleans around noon indicating that this storm would probably be felt in Norfolk Virginia by September 6th and reach the New England states by the 7th. This HC, or more precisely the steering currents, had other ideas and set a course for Galveston.

The Storm Comes Ashore. Then the storm hit on September 8th. I would suggest that no one—who lived on the coast or on an inlet bay—and who had read this book would ride out a major hurricane. Larson painted a frightening story of brick homes and brick schools being torn apart by the rising and violent waters that hit this island with very little warning. Indeed it was the arrogance of the U. S. Weather Service in general, and Isaac Cline, in particular, that essentially preempted any warning. "Cline was one of the era's *new men*, a scientist who believed he knew all there was to know about the motion of clouds and the behavior of storms"[6, 8]. First, the U. S. Weather Service refused to pay any attention to any inputs from Cuba. The U. S. Weather Service had men stationed in Cuba who "said the storm was nothing to worry about." Cuban's "own weather observers, who had pioneered hurricane detection, disagreed." Secondly the U. S. Weather Service insisted that, any storm warning had to come from the Washington office, with absolutely no exceptions. Finally Isaac had the conviction, and public position that no hurricane could ever hit Galveston, as they would be steered north before hand. With such a position Cline had zero motivation to attempt to influence headquarters. However, his "model" of these storms was clearly fatally flawed, or politically influenced, and it resulted in between 6,000 10,000 deaths.

Epilogue to the Galveston Hurricane. Both Cline, and the Weather Service looked arrogant both before the storm, and in defending their overall performance, after the storm. Could Cline have been the first government weather scientist to become defensive and arrogant when questioned on his work? See comments on a Dr. Kenneth Trenberth, below.

Most stories about this storm end on the island. However, this storms destruction was far from over. Indeed the path of this storm, and resulting casualties and damage was almost as terrifying. The storm moved quickly across Galveston Bay, causing severe damage and casualties in neighboring communities, including Houston. Over half the buildings in Houston were damaged[9].This storm continued to move north through Oklahoma, Kansas, Iowa, Minnesota then east to Wisconsin. Torrential rains fell on these states, and six loggers were killed on the Eau Claire River, in Wisconsin. Winds of 80 mph were reported at Chicago on the 11th. This storm refused to die and continued across Michigan into Canada and on to the Maritime provinces where it re-strengthened and killed again: 80 to 100 Canadians were killed, plus an additional 25 -50 casualties in Newfoundland, and 120 on the nearby tiny French islands of St. Pierre and Miquelon. Finally this stormed raced across the Atlantic to Scandinavia and died out over Siberia.

Back on the 7th the Weather Bureau had called for this storm to veer east, over Florida, then up the Atlantic coast, then out to sea. How wrong can one be? Somehow these weather forecasters surely didn't read the steering currents very well.

• The Long Island Express.

This storm started innocently enough, also near the Cape Verde Islands. It moved westerly to north of Puerto Rico, then veered north[10]. Ship reports indicated it was a major hurricane with wind estimates as high as 160 mph. However, Washington forecasters assumed it would pass out to sea, brushing the coast with some wind.

By Cape Hatteras information was available that would suggest the storm was not heading out, but moving due north. Additional inputs were available namely a low pressure system over the Great Lakes, and a high pressure system over the Maritimes. Both of these would contribute to the steering currents guiding the storm north. Again this information was ignored by the Weather Bureau. Tides were also at their highest when both the sun and the moon's gravity tug at the sea. Admittedly even the best forecaster would likely miss the speed of this storm—60 mph—twice the normal forward speed. One junior forecaster in the Weather Bureau did predict this storm correctly, but he was overruled by the chief forecaster, and the general public never saw it coming [10]. This is a performance record remarkably similar to the Galveston tragedy.

This storm cut across the far eastern end of Long Island, then moved onto New England. Rhode Island saw the greatest destruction[11] with the highest water recorded in downtown Providence. "It was storm surge, rising as it was funneled up the ever narrowing bay, which was to prove deadly." Rhode Island deaths reached 390. One report, not verified, claimed that this storm pushed a wall of water 35 feet high ahead of it, sweeping away what ever protective barriers existed at that time[12]. This storm ultimately killed over 700.

Needless to say if such a hurricane had come up the Hudson outlet and scored a direct hit on Manhattan, the casualties would have been analogous to the Galveston hurricane, or worse.

• **Florida Hurricanes.**

The recent HC/TS history in Florida provides a rather dramatic picture of the fickleness of these storms.

In 2005, Katrina passed through South Florida; Rita swept through the Keys and Wilma, after it hit Mexico, re-crossed the Gulf then crossed Florida from Naples to Palm Beach. These three storms were on the top ten costliest list, albeit Katrina and Rita did ~99% of their damage in Louisiana and Texas. More on Katrina shortly.

In 2004, Florida saw massive destruction when it was hit by four major storms. Charley, Frances, Ivan and Jeanne were all on the top ten list and cost the state economy $42 Billion. Note, however, that Ivan hit Alabama first and is therefore classified as an Alabama storm.

Four minor HCs hit Florida between 1992 and 2004, and one major HC, Opal, in 1995. Then Andrew hit in 1992 as a category 5. This was one of the costliest weather disasters ever.

In an interview[13] with William Gray—a well know HC guru, whose credentials will be noted shortly—one question was: "A few years ago when there were quite a few light seasons in a row, you said Florida had just been lucky and it was going to end." The answer: "They've been extremely lucky. The last major hurricane to come through Florida, before Hurricane Andrew hit in 1992 was Hurricane Betsy in 1965, which went through the Keys." "Eight of the last 10 years have been very active - - -." " - - -and yet we went from 1992 until last year with no hurricanes coming through Florida." During this period Gray commented that Florida had "just been lucky and that it was going to end."

Another question: There was a lot of devastation last year [2004]. That doesn't seem very lucky?" "Although last year was a terrible year for them, it could have been worse because none of the four storms that affected the Florida region went into a highly populated area."

• **Mitch**

This HC formed on October 21, 1998, in the southwest Caribbean, and strengthened to a Category-5 HC as it rapidly closed in on Honduras. Now, as a side comment, I had long felt the world would never see Galveston like casualties ever again. How wrong can one get. By October 29 this storm was downgraded from a HC as it approached land. Perhaps this was an unfortunate event, as the steering currents, that Isaac Cline was so proud of, stopped working for several days. And Mitch remained offshore from Honduras and Nicaragua, and *pumped quadrillions* of gallons of water onto the mountains of this region. Some areas received 70 inches of rain over the duration of this storm. The floods and mud-slides killed over 10,000.

• **Katrina**.

The August/September 2005 storm, Katrina, has received incredible press coverage from every possible direction. A Google search on May 27th, 2008 gave the following hits:
· Hurricanes - 25.4 M;
· Hurricanes & Katrina - 12.2 M;
· Katrina - 52.5 M.

In the first 50 citations under Katrina, 48 were on this hurricane. Clearly it will be very difficult to add much new. Here I will focus on the two key legacies for Katrina:
· the storm surge and
· the levee failures.

These were the two dominant causes for destruction and death. More will be added below on the possibility of a couple between HCs and GW.

Katrina left Florida as a Category 1 storm. As it crossed the Gulf it grew to a Category 5, but dropped to a 4, then a 3 as it came ashore, with winds around 120 mph. At 4PM on August 28th, a NHC Bulletin warned of a possible storm surge as high as 28 feet and stressed "some levees could be overtopped." At 4AM on August 29th this storm was 90 miles south of New Orleans. A NOAA buoy reported [14] at 6AM, a "peak significant wave height of 55 feet", the highest ever measured by a government buoy.

Once ashore the resultant tidal surge was perhaps unprecedented, with values of 20 to 30 plus feet being reported. Even though this HC weakened before landfall, several factors contributed to the extreme storm surge:
· the massive size of this storm;
· the strength of this system prior to landfall;
· the very low central pressure at landfall at 920 mb and
· the shallow offshore waters.

The other legacy for Katrina was the failure of the levees. As noted above some levees could fail. And they did. This area surely was warned over and over again that this was going to happen, but no politician was prepared to fight for what was needed. In 2000, one writer[15] suggested the city should change it's nickname from the *Big Easy* to the *Big Worry*.

The levee holding Lake Pontchartrain back was 15 feet high, providing protection from Category 1 and 2 HCs and a fast moving level 3 storm. The levee holding the Mississippi back was 20 feet high. In any event a slow moving Category 3 or any Category 4 or 5 storm, passing within 20 to 30 miles of New Orleans, would be devastating[15].

A report[16]—just after Katrina had hit, and the magnitude of this disaster had become apparent—summarized the situation rather succinctly. "In the wake of hurricane Betsy, 40 years ago, Congress approved a massive hurricane barrier - - - ." "But this project, signed into law by President Johnson, was derailed in 1977 by an environmental lawsuit." The project was estimated to cost ~85M$ in 1965, or ~500M$ in 2005 dollars. In contrast, estimates of the cost of Katrina' damage and reconstruction exceed 100B$.

In any event this project was stopped on December 30, 1977 by a judge who declared the environmental impact statement by the Corps of Engineers "had failed to satisfy federal environmental laws." Wow. Talk about people who "can't handle the truth."

• Gustav

We also evacuated to Lake Travis for HC Gustav, as it rapidly strengthened—to a Category 4, and almost to a Category 5—as it left the west tip of Cuba and moved into the Gulf. Gustav saw the largest evacuation in U. S. history, as more than 3 million people fled this storm. It came ashore on September 1st, 2008 on the central Louisiana coast. Since this storm weakened rapidly and veered to the east, we only spent one night away. However, it still was a very nasty storm causing 153 deaths, with 77 in Haiti and 53 in the United States.

• Ike

We also evacuated to Lake Travis for HC Ike, for four nights, starting on September 10th, 2008. Our son, Brent, decided to ride out Ike and did rather superb work in boarding up our home, plus helping others do the same. We probably should have stayed at Lake Travis for several more nights, for when we returned our power and TV were out for 5-6 more days. Fortunately no damage to our property. Three mid size, branches came down, but only one hit our roof directly, with no damage. Our daughter, Brenda, was not so lucky. She and her neighbor had a four foot diameter tree come down that knocked out three fences plus hit her roof a glancing blow. However, it was enough of a blow to warrant a new roof. And to add to her misery her power and TV were out for three weeks. Needless to say she spent much time at our place. And it also took over a year to get the roof completely rebuilt.

HC Ike was rather remarkably similar to HC Galveston. Both started far to the east. Both traversed over much of Cuba. Both entered the Gulf in early September. Both had Galveston in their sites, albeit with a slightly different pathway. And both did massive damage. Indeed Ike was the third most costliest storm after HC Andrew in 1992 and HC Katrine in 2005.

Unlike the Galveston storm Ike caused 195 deaths, with 74 of these in Haiti. This poor island was hit by four bad storms in 2008, and by a massive earthquake in 2010. The quote by Porfirio Diaz about Mexico: "So far from God - - - ." surely applies to this island too.

Perhaps the major distinction for HC Ike was its huge diameter, over 900 miles. It appeared that "Ike was absorbing and distributing energy over a large area, rather than concentrating it near the center." By September 11th, Ike became the largest Atlantic HC in recorded history.

It finally made landfall early on the 13th. After landfall Ike began a slow turn to the north, and very closely followed the track of the Galveston storm a century or so earlier.

The *Distinguished Veterans*

Recently Dr. Kenneth Trenberth has accused the skeptics of not listening to him when he states that all weather events are influenced by global warming. He claims he knows the truth. Trenberth is not the only *warmer* to argue that HCs are caused by GW. For the layman it is almost impossible to sort out who knows the *truth* when two sets of scientists disagree. I believe there is a way around this dilemma, and that is to study the views of what I call the *distinguished veterans – the DVs*. These *veterans* are scientists with rather impressive credentials and accomplishments. Most of them are retired. These individuals do not have to play the game of chasing after grant money. These are scientists that do not have to curry favor with the department chair-person, or other university-institute-administration *brass*. They are free to state their convictions.

Now any input or quote or even a scientific paper can be biased. This *contamination* can be due to a political agenda or to career concerns. A recourse to the views of the *DVs* would eliminate the career concerns factor, and may also go a long way to eliminating the political agenda factor.

What follows then are the views from four *DVs* on GW, HCs and the possible couple.

• Dr. Neil Frank.

This review will start with Dr. Frank, Chief Meteorologist for Channel 11 in Houston, from 1987 to 2008. Although he was former director of the NHC for 13 years, with a total of 26 years spent at that Center, and he started his professional career as a weather officer in the Air Force, however, he has not been an active scientist in weather or climate research.. Nonetheless he is very knowledgeable on this subject. I have heard him speak on the GW issue several times at various professional and service meetings.

His concerns include an over reliance, by the *warmers*, on computer models. Dr. Frank is highly skeptical about the global warming issue. He is convinced that there is something very wrong with the proponents case. He has noted[17,18] that the models, used in weather forecasting can't be relied on for a three day forecast. He also notes, that for global warming predictions, we are being asked to rely on similar, but simpler models than those used in weather forecasting. Yet these climate models are applied in a far more complex arena to prepare climate projections for 50 to 200 years.

• Dr. William Gray.

Gray obtained his BSc degree in 1952, then worked for the Air Force forecasting weather. He attended U. of Chicago from 1957 to 1961, obtaining an MSc degree in Meteorology in 1959, and a PhD in Geophysical Sciences in 1964. He joined the Department of Atmospheric Sciences, at Colorado State U., in 1961.

Gray has been forecasting the annual number of hurricanes for many years. He has become the nations most prominent hurricane authority. He is a Fellow of the American Meteorological Society and a recipient of many awards.

Gray is a long-term skeptic the global warming issue. He is most out-spoken and perhaps the most controversial scientist on this issue. His statements are aimed at putting the computer models used in this field into perspective. Key quotes:

· In a speech in Houston[19] he noted that *climate models, while surely useful, are far from perfect. "The models have been superb when used for the next 5-10 days, but when modelers move out onto the climate area the complexity becomes too damm much."*

· Some have tried to couple the upswing in hurricane activity as evidence of global warming. Gray's reaction[20] to such claims: there is no way such an interpretation can be accepted. Anthropogenic greenhouse gas warming, if a physical valid hypothesis, is a very slow and gradual process that, at best, could only be expected to bring about small changes in global circulation over periods of 50 - 100 years. This would not result in abrupt and dramatic upturn in hurricane activity.

· While Gray is in *retirement* he still works every day. "For years I haven't had any NOAA, NASA or Navy money. But I'm having more fun." - - - "Right now I'm trying to work on this human-induced global-warming thing that I think is grossly exaggerated." He admits he has cut his forecasting project way back, partly do to lack of funding. He noted he had "NOAA money for 30 some years, and then when the Clinton administration came in, and Gore started directing some of the environmental stuff, I was cut off. I couldn't get any NOAA money. They turned down 13 straight proposals from me."

· "Nearly all my colleagues, who have been around for 40 to 50 years, are skeptical as hell about this whole global warming thing."

• Dr. James O'Brien

This *DV* has a BSc in Chemistry from Rutgers and an MSc and a PhD in Meteorology from Texas A&M. He has worked for DuPont, the Air Force, the University of Colorado, Oregon State University and Florida State University, where he is the Distinguished Professor of Meteorology & Oceanography. He has also served as the Florida State Climatologist.

O'Brien is credited[22] with ocean modeling which led to new breakthroughs in understanding coastal up-welling, El Nino, La Nina and HC effects on the ocean. He is probably best known for his basic research[23] into the El Nino phenomenon.

Some recent comments follow.

· I don't know of anybody who would think that global warming is causing Katrina.

· There is no correlation between HC intensity and GW. Rather there seems to be about a15 year building and a 15 year waning cycle.

· Many have argued that, with GW, the ocean temperatures have gone up. O'Brien commented[24] that while the ocean may be warming in spots, it is not so in the HC formation part of the Atlantic. From 5°N to 20°N and from Africa over to near the United States, it has cooled.

• Dr. Tad Murty.

This *DV* is from India, with a PhD from the U. of Chicago in Oceanography/Meteorology.

Murty was a research scientist for the Canadian Fisheries and Oceans for 27 years. He was Canada's representative in the design of the International Tsunami Monitoriing System. He also has served as Director of the Australian National Tidal Facility for 3 years. In his spare time Dr. Murty has served as editor of the International Journal of Natural Hazards for 18 years. Murty has now returned to Canada and is affiliated with Carleton University and the University of Ottawa.

Murty is very well known from his work on tsunamis. He is an expert on tsunamis and other ocean phenomena, and the former president of the Tsunami Society. After the devastating tsunami of December 26, 2004, Murty fielded[25] more than 400 media interview requests, in the first half of 2005 alone.

Over his career, Murty has made many comments on global warming and on hurricanes.

· "Out of 2000 or so manuscripts that crossed my desk in my 18 years as editor of the International Journal Natural Hazards, I cant recall a single one based on actual observations that claimed that global warming has anything to do with extreme weather events [26]."

· "Yes many papers on computer models tell a different story. "But after being associated with such simulations for the past 45 years, I have little faith in their predictions. "With a very slight tweaking of one single parameter (low cloud cover) in the model, forecasts can change abruptly from global warming to an ice age."

· He recently commented that if human activities are increasing the frequency and intensity of hurricanes it is not obvious in the observational records, through 2004. Out of 20 atmospheric and oceanographic parameters associated with hurricanes, not a single record was set after October 1979.

· "The two basins in the world most impacted by hurricanes are the Bay of Bengal in South Asia and the Gulf of Mexico. "Since 1995 there has been an increase in the annual number of tropical cyclones in the Gulf of Mexico. "However no new records have been set and nothing that cant be attributed to natural variability is happening."

· "My colleagues in India and I put together a 200 year database[26] and found that the total number of cyclones in the 20th century is about half of that in the 19th. "For the state of Onissa, which is the most affected in India, there were 72 storm surges in the 19th century and 56 in the 20th. "There were three super cyclones in the 19th century and only ine in the 20th."

· The next quote focuses directly on Murty's view about global warming. "This is the biggest scientific hoax being perpetrated on humanity [27]. "There is no global warming due to human anthropogenic activities. "The atmosphere has changed much in 280 million years and there have always been cycles of warming and cooling."

Hurricanes and Global Warming

• Introduction.

Today one hears that global warming is the culprit behind such intense storms as Katrina and Rita. Many Gorons have frequently expounded on this idea. And some of these are government scientists. For example Dr. Kevin Trenberth was recently cited by Time magazine. I am aware of Trenberth from the work I did on[28] the huge computer models used to simulate the climate. He declared then: "that the burden of proof, that a model result is not valid, should be on the critic and not the modeler." This position is exactly opposite to what is standard practice for models developed in industry or by consultants. I can imagine the reception I would get if I had that arrogant attitude with the management of the two companies I had worked at.

In any event many high horsepower scientists have debated this issue, even before the Katrina storm. Two main questions have been debated namely if global warming is causing an increase in:

(a) the frequency of TS and HCs? and

(b) the ferocity of Hcs?

• Preliminary Debates over this potential couple.

The debate over this potential couple has been long, complicated and convoluted. Those interested in details on this debate are referred to NOTE 9.

Conclusions

I have briefly reviewed my credentials here to be an objective and accurate observer of the hurricane scene. I have reviewed many statistics on this subject. I have summarized the stories of five major HCs—Galveston, the Long Island Express, Mitch, Katrina and Ike—and commented briefly on others. I have reviewed the views of four *Distinguished Veterans* in the HC field. And I have reviewed the arguments on both sides of this debate on the potential couple between GW and HCs.

Now the average citizen may still feel uncomfortable with my ability to give the true scoop on these storms. For the layman it is difficult to sort out who knows the *truth* particularly when two sets of scientists disagree. I believe there is a way around this dilemma, at least in part, and that is to study the views of what I call the *distinguished veterans*. The italics are used to distinguish these scientific veterans from the military. The *veterans* discussed here are scientists with impressive credentials and accomplishments. Most of them are retired. These individuals do not have to play the game of chasing after grant money. These are scientists that do not have to curry favor with the department chair-person, or other university/agency/institute *brass*. These *DVs* are free to state their convictions. I have studied the history of over 60 *DVs*. And the vast majority of these are all highly skeptical on the AGW issue. Brief reports from four *veterans* were studied, all experts on HCs.

There is a serious debate on global warming underway, and the skeptics, including the views of the *DVs*, may very well have the best of the arguments. To claim that these scientists don't understand the sciences involved on global warming and hurricanes, is silly. To argue that these *veterans* are being conned into their positions is spurious. And to accuse them of being puppets of the energy industry is specious at best.

In closing it is important to note that one still needs to plan and respond to imminent HCs as they develop. Don't ignore. Don't underestimate, but do not worry if GW was the cause.

References and Notes

(0) *The Saffir-Simpson Hurricane Wind Scale*, Feb.27. 2010. http://www.nhc.noaa.gov/sshws.shtml.
(1.1) The title for this chapter is taken form the movie *My Fair Lady*.

(1.2) *Hurricanes, Typhoons and Cyclones.* Undated.
See h ttp://www.econet.org.uk/weather/hurri.html.

(2) *Tropical Storm Claudette (1979)*, Wikipedia. This source also noted that this storm name was not retired and has been used in 1985, 1991, 1997 and 2003. It is scheduled for use again in 2009.

(3) *The Great Flood of 2001*, Texas - Houston Chronicle Magazine, July 15, 2001.

(4) *Hurricane Alicia, 1983*, USA Today, August 30, 1999.

(5) Westbrook, G., Papers at Offshore Technology Conference (OTC) meetings, Houston, Texas

(5.1) *The Incredible Story of the World's Oceans: Will GW Have an Impact*, #8689, May 4, 1998.

(5.2) *Global Warming and the World's Oceans - Update*, #10773, May 3, 1999.

(5.3) *Global Warming and the World's Oceans - The Millennium Outlook*, #12115, May 5, 2000.

(5.4) *The Global Warming Issue: Current Status*, #14284, May 6, 2002.

(6) Larson, Erik, *Isaac's Storm*, , Crown Publishers, New York, 1999.

(6.1) See NOTE 8.

(6.2) See p80 for inputs on Isaac's 1891 article where he claims it was essentially impossible for a hurricane to hit Galveston.

(7) MacNelly, Jeff, *Shoe*, Houston Chronicle, May 22. 2006.

(8) *Author recounts 'the deadliest hurricane in history'*, CNN book news, August 25, 1999.

(9) Heidorn, Kieth, The 1900 Galveston Hurricane, The Weather Doctor, September 1, 2000. See:
http://www.islandnet.com/~see/weather/events/1900hurr.htm.

(10) Cotterly, W., Hurricanes & Tropical Storms: Impact on Maine and Androscoggin County, 1996

(11) *The Hurricane of '38*, American Experience, © 1999 - 2002. See:
http://www.pbs.org/wgbh/annex/hurricane38/maps/index.html.

(12) *New York's Worst Hurricane Fears Confirmed in New Study*, Environment News Service, October 28, 2006. See: http://www.ens-newswire.com/ens/oct2006/2006-10-26-01.asp.

(13) Svitil, Kathy A., *Discover Dialogue: Meteorologist William Gray*, V26, N9, Discover, September, 2005.

(14) Graumann, Axel, et al, *Hurricane Katrina - A Climatological Perspective - Preliminary Report*, Technical Report 2005-01, NOAA'S National Climatic Data Center, October, 2005.

(15) West, James and Vaccaro, Chris, *'Big Easy' a bowl of trouble in hurricanes*, USA Today, originally published in July 2000 and updated on August 28,2005.

(16) Vastabedian, Ralph and Pae, Peter, *A Barrier That Could Have Been*, LA Times, Sept. 9, 2005.

(17) See Houston Chronicle, April 11, 1998 as reported by the South Florida Sun Sentinal.

(18) See Richmond Times Dispatch, April 11, 1998. This was also reported on the Internet at www. junkscience.com/news2/richtime.htm.

(19) Gray, W., Colorado State University, *Predicted Hurricane Activity for 1997: Is Global Warming Causing More and Bigger Hurricanes?*, Speech at the NHA meeting, Houston, TX, April 25, 1997.

(20) Gray, W., et al, *Early April Forecast of Atlantic Basin Seasonal Hurricane Activity for 1997*, Dept of Atmospheric Science, Colorado State University, April 4, 1997.

(21) Not assigned.

(22) Campbell, Ellen, *Cambridge Who's Who names James O'brien Professional of the Year in Meteorology*, April 18, 2007.

(23) O'brien, James, *Atlantic Hurricanes: The True Story*, Washington Roundtable on Science & Public Policy, George C. Marshall Institute, October 12, 2005.

(24) Glassman, James K., *Hurricanes and Global Warming: Interview with Dr. James J. O'Brien*, Capitalism magazine, September 13, 2005.

(25) *Telugu Tad Murty Wins Indo Canada Award, June 11, 2005.* See: http://www.tlca.com.

(26) Murty Tad, *Katrina and History*, Montreal Gazette, September 1, 2005.

(27) List of scientists opposing the mainstream scientific assessment of global warming, Wikipedia. This list covers yhe views of dozens of high horsepower skeptics, including Murty. Latest reference used was dated November, 30, 2009.

(28) *When Models and Satellites Mislead*, NCAR, March 13, 1997. This report covers two papers published in Nature of the same date. The key quote by Kenneth Trenberth is in the second paper: The use and abuse of climate models.

(29 - 39) See NOTE 9

Chapter 4.3 Sea Level Rise

Background

Previous writings have reported that Florida—in addition to the tidal wave of attention it has received due to its voting systems—has been in the news due to the global warming issue and its alleged impact on hurricanes and sea level rise (SLR).The subject of sea level rise will be examined for each component of sea level change. The Antarctic has been viewed by many to represent the component with the greatest risk from global warming. If such views proved to be correct, catastrophic sea level rise would be likely. Hence the cryosphere will be covered in some depth in this chapter, with effort made to separate the science from the hype that engulfs this subject.

Several themes will be noted in this background report.

Awesome nature of natural events.

• Noah's Flood?

One of the more dramatic events[1] associated with SLR occurred between the Mediterranean Sea and the Black Sea. As the world emerged from the last ice age, from about 17 Kiloyears before present (KYBP) to about 7.5 KYBP dramatic changes occurred in the level of the worlds oceans. The Mediterranean was connected to the Atlantic and its level gradually rose from 110 meters below present (mbp) to about 15 mbp. However, the Black Sea was then isolated from the Mediterranean Sea and its level was controlled by runoff from Russia and east Europe and by evaporation. Around 7.5KYBP a huge hydrostatic head of over 460 feet developed between the two seas. Leakage started and gradually increased to a major flow then soon a gigantic flood, with the Black Sea soon down to the level of the Mediterranean. Consider the following tabulation.

Item	17 kybp	14 kybp	9 kybp	7.5 kybp	7.499 kybp
Med. Sea level , mbp	110	90	30	15	15
Black Sea level , mbp	140	15	127	156	15
Head, meters	-	-	-	141	0
Head, feet	-	-	-	463	0

Mediterranean & Black Sea Level Changes in Coming out of the Last Ice Age.

More on this subject of Noah's Floods will follow shortly.

• Astrophysical Phenomena.

As with any warming it is desirable to separate natural warming from any anthropogenic warming. The acronym DNA is useful here, namely one must strive to Distinguish Natural from Anthropogenic warming. The same is true with SLR. In consideration of natural warming the solar interactions with our climate may be the number one factor. Indeed a recent article[2] listed 14 areas of a solar-climatic couple. Two of these follow

· The Aurora Borealis - On March 6, 1716, the Aurora Borealis, also known as the Northern Lights, returned to England. Few, if any, had ever seen this phenomenon of *The Ghost Riders in the Sky* as it had been absent from these skies for over 70 years, since 1645 AD. "Frightened servants thronged the street convinced that the day of judgement had arrived[3]." Scientists now know this was the end of the Maunder Minimum, a period where sunspots on the surface of the sun virtually disappeared.

Sun-spots are a freckle like phenomenon that appear on the surface of the sun, over a typical 11 year Solar cycle. Sun-spots are believed to represent areas of localized magnetic activity on the sun. They represent one of the attributes of the sun that can be observed and measured from Earth. As such, they represent an indirect measure of the activity of our sun. Proponents, of the anthropogenic basis for global warming, frequently mention that the carbon dioxide concentration in our atmosphere has increased by about 30 % over the past century. In contrast they never mention that this solar magnetic field has more than doubled[4] over the past 100 ore so years.

Without these sunspots and the magnetic activity and the solar wind that comes with them, the Northern Lights disappeared.

This 11 year cycle modestly affects the solar output. Supporters of the AGW theory have repeatedly argued that this variation in solar output is just too small to have any impact on our climate. However recent data[5a] suggest that galactic cosmic rays (GCR), rays that bombard the Earth continuously, are modulated by this solar magnetic cycle. This increases or decreases their assistance in cloud forming activity, in counter synchronization with the solar cycle. This process, along with variation in solar output, together could be sufficient to explain most of the short term temperature variation we have seen this century.

· The Solar Wind. Besides sunspots and solar radiation, there are a variety of other solar attributes of import: solar flares, corona maximum ejections, the solar wind, the overall magnetic flux within the sun and finally the UV part of the solar spectrum. Perhaps the most spectacular phenomenon is the flow of charged solar particles[5b]—the Solar Wind—around Earth. This is controlled by the Earth's magnetic field. And research has indicated that since this activity is clearly coupled to sunspot activity and to the strength of the solar radiation hitting Earth, this flux could also be involved in the warming of our planet.

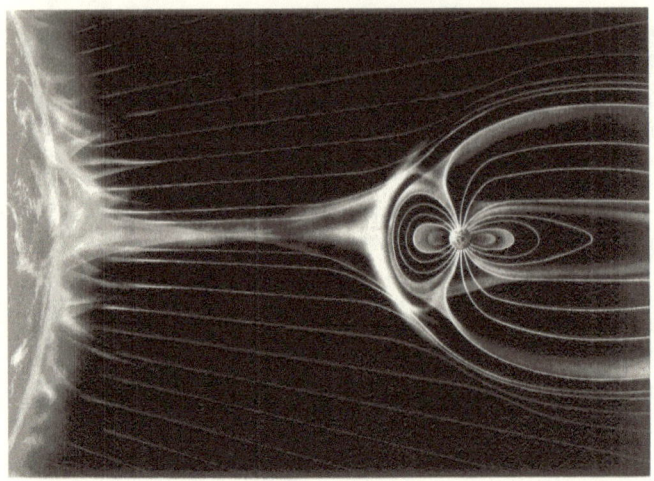

A Depiction of the Solar Wind from Sun to and around Earth

The sun's magnetic field controls the flux rate of GCRs into the inner solar system where Earth resides. As GCRs drill down into our lower atmosphere over the oceans they ionize particles of moisture to form clouds. Greater cloud cover causes sunlight to be reflected back into deep space. When there is less sunlight, the Earth cools. This is a natural process. This process, related to changes in the sun's magnetic field, modulates Earth's temperatures.

It is surely premature to claim that the *science is done* on this subject. However, it's claims received a major boost via lab work by Henrik Svensmark at the Danish National Space Center in Copenhagen, where he demonstrated[6] the process in the World's largest cloud chamber.

See additional key references[7, 8] on the work by Svensmark and his associates for more details. See also another key reference[9] for 14 citations on this subject from a wide spectrum of scientists

In summary astrophysicists see a multiplicity of evidence - from hard data to educated assessments - that our climate is impacted by solar, stellar and galactic phenomena.

• Climate Science is in its infancy.

It must be fairly obvious that our understanding of the Solar magnetic flux, the behavior of the Sun's heliosphere, the Earth's magnetosphere and GCRs are at the *embryonic* end of the investigation spectrum, not *the science is done* end. The emergence of astrophysics, as one of the key global warming sciences has been one of the major developments of the past 10 - 20 years. Further, this theme of incomplete science is not limited to the astrophysics field, but is spread throughout all the major sciences encountered in this field. Surely this is true on the hydrosphere and such subjects as ocean circulations, ocean temperatures and sea levels.

Measurements of Sea Level

A quick Google inquiry on "sea level rise" yielded 2½ million citations. The reference[9b] covered nine of these, with publication dates from 2001 to 2007. More on these shortly.

Today, one hears much about the impending disasters that loom around the world due to catastrophic SLR. However, seldom does one hear much talk about the task of measuring SL and hence SLR.. One has to deal with tides, winds, changing air pressures and even crustal rebound. It is difficult enough to measure these at one location let alone in enough spots around the world to come up with a meaningful average. Several inputs on SL measurements follow.

• **Sea Level Rise - Coming Out of the Last Ice** Age
The geological record[10] shows the sea level has risen by 120 meters over the past 20 Kilo years (KYs). This rise was non linear. Values ranged from 2 to 12 mm/year. A recent graph[11a] of estimated sea level rise (SLR) rates, on a 250 year time unit, showed a maximum rate of about 18 mm/year. In addition there are undoubtedly major spikes in the rise rate that occurred due to the existence of what are termed melt water pulses. An example of such an event would be when ancient Lake Missoula in Idaho and Montana, a lake the size of Lake Erie, and formed by an ice dam, gave way about 16 KYs ago. When such an event occurred the SLR rates would likely be higher for a short period of time.

Another example[11b] can be seen in the Glacial Lake Agazziz's (GLA)—a lake that covered parts of the Dakotas, Minnesota, Manitoba, Saskatchewan, Ontario and Quebec—catastrophic collapse, when gigantic volumes of water poured into Hudson Bay. Coastal dwellers would see some severe flooding, with those in the Persian Gulf region the most impacted. During the Last Ice Age, with sea levels down by as much as 120 meters (394 feet), this region would be dry, but with many inhabitants then living on that land. While the melt water pulse from GLA would have caused a rapid SLR of about 50 centimeters, this would be sufficient to flood parts of the dry Persian Gulf area for many kilometers. As such this flood may well have been the basis for the epic stories of Gilgamesh, Sumerian and Mesopotamian flood legends. And Noah's Flood may not have been in the Black Sea, but a bit further south.

• Sea Level Rise - Changes over the past century

For the contemporary period some reports have claimed recent rates as high as 8 mm/year, but such rates have proven to be erroneous. Still other reports have argued the rate has been no more than 1 mm/year over the last 200 years. The1995 UN report[12] —The Intergovernmental Panel on Climate Change (IPCC) report—provides a reasonable perspective, with a range of sea level rise rate observations of 1 to 2½ mm/year over the past century. The 2006 citation[9b] reported a trend of 2.4 mm/year over the period 1993 - 2000. It also reaffirmed the notion that this is a non-linear phenomenon.

The measurement of sea level is confounded by several factors, as noted above. Crustal rebound is still underway from the compression applied over the Last Ice Age. This has to be isolated from measurements taken from tidal gauges. And the changes in air pressure, such as the El Niño events, can lead to water being pushed substantially higher in parts of the Pacific. Again this must be taken into account.

The uncertainty in sea level rise over the past century has been reviewed[12, 13] for both the observed rise and the estimated rise. For the observed rise we see an uncertainty range of 15 cms. This is based on estimates for low, middle and high values of 10, 18 and 25 cms.

The authors of the UN report ask: can this sea level rise over the past century be explained? They give estimates for Low, middle and high values for each of the components of sea level rise follow, along with comments by the UN in italics.

Item	Low - cms	Med. – cms	High - cms	Range - cms
Observed rise	10	18	25	15
Estimated rises				
1. Thermal expansion	2	4	7	5
They noted that observational data are still too sparse to make global scale estimates.				
2. Changes in surface/ground water	-5	0.5	7	12
Estimates are uncertain/speculative.				
3. Glaciers/small ice caps	2	3.5	5	3
While many of the worlds glaciers have retreated, there are mixed signals in the data. Continuous long term measures of the mass balances are very limited.				
4. Greenland Ice-sheet	-4	0	4	8
There is insufficient evidence from the models and data.				
5. Antarctic Ice-sheet	-14	0	14	28
There is insufficient evidence from the models and data.				
Total estimated rise	-19	8	37	56

UN Estimates of Observed SLR and Estimated SLR - cms

Note that there is a range of 56 cms between the total low and high estimates. Also note that the zero values above should be interpreted as a measure of the poor state of knowledge in 1996.

Clearly there is still uncertainty, both in our ability to measure these recent sea level changes, and also in our ability to analyze the situation and estimate these changes. Here the range in observed estimates was 15 cms. Hence, the estimate uncertainty is nearly 4X greater than the observed uncertainty.

• Sea Level Rise - The next 100 years

The huge uncertainty seen in the above SLR estimates does not bode well for our ability to forecast the future. Just as global warming forecasts for temperature increases have fallen over time, so have those for sea rise. Forecasts have declined since the early 1980's when a 7 to 8 meter rise was cited (for the year 2130). This level fell to less than 1 meter by 1990 and to less than a half a meter (49 cms) in 1995, a value that is clearly highly uncertain, in 1995. This is now down to as low as 21 cms today. A key EPA forecast[13] of future SLR comes in at 34 cm. The breakdown of these two forecasts is noted below, in centimeters.

Item	UN – cms	EPA - cms
Thermal Expansion	28	21
Glaciers/Small ice caps	16	9
Greenland ice-sheet	6	5
Antarctic ice-sheet	-1	-1
Total estimated rise	49	34

Comparison between UN and EPA Forecasts - cms

This exercise suffers not only from a weak understanding of the natural processes involved, but also from a less than perfect picture of the likely future global warming.

A comparison is made below between the maximum rise, due to the complete melting of the cryosphere, against the current UN forecasts. Such forecasts have been based on a scenario of the future where the warming is set at 2½ ºC, a level many skeptics would argue is too high. However, SLR contributions for the three components under this scenario follow.

Item	SLR with 100% melt, meters	SLR forecast
Antarctic ice-sheet	76.0	-1
Greenland ice-sheet	7.6	6
Glaciers/Small ice caps	0.5	16
Totals	84.1	21

Comparison of the "Armageddon Scenario" versus current UN Forecasts.

Bringing these together we see that the UN forecast for these three components, for the next century, is 21 cms versus the ultimate potential of 84.1 meters. This is **less than ¼ of one percent** of the ultimate, hardly an apocalyptic outlook.

• Variation of Sea Level rise Around the Globe

As one scans the globe one sees there are many areas that are more vulnerable to SLR than other areas. For the United States, maps[13b] have been developed that outline such situations. The writers describe 1.5 and 3.5 meter contours that represent the areas inundated. However, for a variety of reasons the 1.5 meter contour represents the area inundated by a 70 cm rise. This is close to the 49 cm forecast noted above. However such maps permit government officials to develop plans to adapt to this potential future.

Now some government scientists are starting to make the case that the situation is even more complicated than as described above. They note[14]: "as the world considers how to adapt to 21st-century climate change, researchers are acknowledging that a single value of projected sea level doesn't fit all." They argue that trends in temperature and salinity are not unfolding equally around the globe. Hence sea level is not a physical constant.

In contrast one of the papers cited [9b] earlier, reported [15] an analysis of regional SLR, over the 1950 to 2000 period, and concluded "that it is not possible to detect a significant sea level rise over this period anywhere." Just as it is with the attempts to take GCMs based temperature predictions regional—where there is serious concerns if such models are valid on a global scale and there will be even more concerns on a regional scale— so it is with SLR.

Hype, Hype and more Hype

Our former VP has received rather incredible coverage on his second book: *An Inconvenient Truth(AIT)*. See Reference 1, §2.4. However, I approached *AIT* believing it would be loaded with propaganda and hype. I wasn't disappointed. I see this in several ways, but particularly in his selection and coloring of many high quality colored pictures, graphs and maps. However, this is part of his propaganda pitch. For example Gore shows three Greenland pictures/maps with snow–ice cover shown in white and the surface meltwater in very bright red, a color that surely distorts the situation. The conclusion, that the uninitiated will draw, is one of imminent and terrifying collapse of this ice sheet and subsequent sea level rise. However, Gore does not give the reader any sense of the incredibly low probability for such an outcome. As such Gore is the leader in the use of propaganda /hype.

We will examine the use of propaganda or hype on three subjects related to SLR: Antarctica, the Himalayan glaciers and the retraction of a Key Report on SLR.

• **Antarctica**

(1) <u>Background</u>. Two areas are singled out to show this distortion. These are the Larsen Ice-Shelves and the West Antarctic Ice-Sheet (WAIS). Note the definitions for an **ice-shelf** and an **ice-sheet.**

· An <u>ice-sheet</u> represents the ice reservoir over land, representing over 97% of the ice volume.

· An <u>ice-shelf</u> is ice floating at the edge of an ice-sheet. Further, melting of an ice-shelf would not contribute to sea level rise as shelves are already displacing an equal volume of water.

(2) <u>The Larsen Ice-Shelf</u>. This is situated next to the Antarctic Peninsula, the land that stretches north towards Chile and Argentina. As an <u>ice-shelf</u>, it is frozen to the <u>ice-sheet</u> on the peninsula.

It should be noted that this peninsula is hardly representative of the conditions in Antarctica. First, it is substantially north of the South Pole, and situated above the Antarctic Circle. Secondly being a peninsula it is tremendously impacted by ocean currents and ocean temperatures. And finally its climate is closer to that of southern Chile, than it is to the bulk of Antarctica. A New York Times writer[16] noted: "this peninsula is a 700 mile long rocky kite trail curving out from the coldest, driest, highest continent on Earth." While he noted these unique geographical characteristics, he did not note it's distinct climate. Indeed he went on to report that this peninsula has "seen a jump of more than 5 degrees in just 50 years, including a 10 degree average warming in its winter years." The implication was made that such temperature changes were due to AGW.

There have been many articles about loss of a piece of this shelf. One example was a report that concluded [17]:

· the peninsula temperature has increased about 2½ degrees C. in the past 50 years;

· the rest of this continent – the other 96% – shows no current indication of rising;

· there is no unusual significant loss of ice of any kind from the larger 96% of Antarctica;

· ice shelves, such as Larsen, break up with rising temperature, but since they are floating they will not affect sea levels, directly.

This reference provided pictures of this ice shelf dated 17th February 2002 and 5th March 2002. This shelf was about 220m thick and during a 35 day period a section of about 3,250 km² split off. It is thought to have been in existence for at least 400 years prior to this and probably as long as 12,000 years, since the end of the last ice age.

One other input, also from the New York Times, was by a noted skeptic[18] of the AGW position. He confirmed that an ice shelf "does not raise sea levels upon melting." Yet the collapse of this ice shelf "has proved to be a perfect natural disaster for 'the Apocalypse Now' school of journalism." He went on: "It is now perfectly clear that we are all doomed and that this is the wake up call for urgent action on greenhouse emissions, automobiles, industry, and virtually everything else to do with economic growth."

(3) The WAIS. A similar story exists on the treatment of **WAIS**. For example, consider the treatment a science magazine[19] gave to a key paper. First the title, then the first sentence:

· title, as published in the J. of Glaciology: *Rapid Sea-Level Rise Soon from WAIS Collapse?*;

· first sentence, as written by the author: *Will worldwide sea level soon rise rapidly because of a shrinkage of the WAIS?*

In short the magazine changed a serious question to a conclusion and changed the word *shrinkage* to *collapse!*

More recently, a paper published in The Geological Society of America [19b] noted: "Even if there is a very low probability, future effects on the stability of the WAIS and associated sea-level rise should not be ignored, as the rapid changes in the past 20 years resulting from global warming could be accelerated by subglacial volcanism." Note that this writer did not cite AGW. This view is similar to the views [19c] of Dr. Judith Curry of Georgia Tech. She is almost a convert to the skeptical side. She has become a middle-of-the-roader. However, some see her now as a heretic. In any event she believes in AGW, but at the low end of range. She is also concerned on the WAIS. It deserves a much higher priority than the UN gives it, and calls it a looming catastrophe that could happen in the next few generations. However, her major concern is "on issues associated with the integrity of climate science."

The writer of this next article noted that the WAIS stability is one of wide spread interest, but of major disparity.[19a] This scientist stated that "it is difficult to see how climate warming (whether anthropogenic or natural) will twigger a collapse of the WAIS in the next century or two." This conclusion echoed an earlier one[20] that stated "The WAIS is not melting rapidly, is reasonably stable and has been so for more than a century."

• Himalayan Glaciers

Recently the IPCC released [21] a benchmark report that was claimed to incorporate the latest and most detailed research into the impact of global warming. This report suggests that the world's glaciers were melting so fast that those in the Himalayas could vanish by 2035. It now turns out that this "detailed research" was based on chance comments, speculation and was not supported by any formal research.

• A Key Report on SLR is Retracted.

Anthony Watts commented [22] on several reports on the expected SLR by 2100. Examples:
· July 2007 - IPCC "sea level will probably rise [22, 23] by 18cm - 59cm."
· July 2009 - Nature Geoscience "sea level would rise [22, 24] by between 7cm and 82cm."
· December 2009 - PNAS "the relationship projects a sea-level rise [22, 25] ranging from 75 to 190cm."
· February 2010 - the July 2009 claims are retracted.

Now it is hard to claim that this input was truly hype. Indeed, the lead author claimed: "it's one of those things that happen. People make mistakes and mistakes happen in science." However, "a formal retraction was necessary, rather than a correction, because the errors undermined the study's conclusion."

Conclusions

The analysis of the above subjects is far from complete. But the assessments to date do not represent any sort of apocalypse. The above material is presented to point out the frequent imbalance between headlines and the facts. This subject of global warming is so complex that the media needs to be more cautious and precise and to place this subject into context. It needs to be on a constant vigil against those who would take advantage of it to further their own political agenda.

There is considerable agreement that the Earth has seen a modest amount of surface warming over the past century, due either to anthropogenic or natural causes. Although our knowledge of this subject is very incomplete there is a reasonable probability that this trend will continue and we should not be surprised at a modest sea level rise. One key question that remains to be answered is what fraction of any SLR is due to natural warming and what is due to anthropogenic warming?

In the meantime areas that may be vulnerable need to be prepared for say a 1, maybe up to a 2 foot rise over the next century. This may mean doing nothing in some areas, to building/extending dykes/sea-walls in other areas. It may mean discouraging sea-side dwellings in some areas to actual relocation of some population in other areas.

References and Notes

(1a) Westbrook, Gerald T., *The Incredible Story of the Worlds Oceans: Will Global Warming Have an Impact*, Offshore Technology Conference, paper #8689, May 4, 1998. Also printed in USAEE Dialogue, **7**, (April 1999).

(1b) Westbrook, G., *Global Warming and the World's Oceans: (a) 1999 Update and (b) The Millennium Outlook*, Offshore Technology Conference, Papers # 10773 and 12115, (May 1999 and 2000).

(2a) Westbrook, Gerald T., *Global Warming: Introduction and Interactions with the World's Oceans*, Offshore, June, 1999.

(2b) Westbrook, Gerald T., *Global Warming and the World's Oceans Part II*, Offshore, Sept., 2002.

(3) Fara, P., *Learning from the Past*. See the Global Warming Debate - The Report of the European Science and Env. Forum, edited by J. Emsley, Bourne Press Ltd, Bournemouth, Dorset, 1996.

(4) Lockwood, M. Et al, *A doubling of the Sun's coronal magnetic field during the past 100 years*, Nature 399, 437-439 June 3, 1999.

(5a) Svensmark, H., and Friis-Christensen, E., *Variation of cosmic ray flux and global cloud coverage ... a missing link in solar-climate relationships*, J. of Atmospheric and Solar-Terrestrial Physics, 59, 1225-1232, 1997.

(5b) From a NASA/Marshall Space Flight Center on Solar Physics, Why We Study the Sun Space Weather. See: http://solarscience.msfc.nasa.gov/Whysolar.shtml.

(6) The Cloud Mystery, EuropeNews, April3, 2010.

(7) Svensmark, H. And Calder, N., *The Chilling Stars*, Totem Books, March 19, 2003.

(8) Watts, Anthony, *Svensmark: "global warming stopped and a cooling is beginning" – "enjoy global warming while it lasts."* See: Watts Up With That? web site, October 9, 2009.

(9a) Khandekar, Madhav L., *Questioning the Global Warming Science: An annotated bibliography of recent peer-reviewed papers, Section 2 - Impact of solar variability on the earth's climate*, Friends of Science, January 2007.

(9b) Khandekar, Madhav L., *Questioning the Global Warming Science: An annotated bibliography of recent peer-reviewed papers, Section 3 - Impact of solar variability on the earth's climate*, Friends of Science, January 2007.

(10) Fairbanks, R. G., *A 17,000 year glacio-elastic sea level record: influence of glacier melting rates on the Younger-Dryas event and deep ocean circulation*, Nature **342**, December 7, 1989.

(11a) McGuire, W. J. Et al, *Correlation between rate of sea level change and frequency of explosive volcanism in the Mediterranean*, Nature **389**, October, 2, 1997.

(11b) Boswell, Randy, *Noah's Flood: the Canada connection. Scientist Links Lake Agassiz To Noah's Flood*, Winnipeg Free Press, May 9, 2004.

(12) IPCC, Climate Change 1995 - The Science of Climate Change", Cambridge U. Press, (1996).

(13a) Titus, J. G. and Narayanan, V. K., *The Prob. of Sea Level Rise*, EPA 230-R95.008 (Oct. 1995).

(13b) Titus, James G., (2004), *Maps that Depict the Business-As-Usual Response to Sea Level Rise in the Decentralized United States of America*, OECD Global Forum on Sustainable Development and Climate Change, ENV/EPOC/GF/SD/RD(2004)9/ Final, OECD, Paris.

(14) Hensen, Bob, *rough Sea regional variations add a wild card to future sea-level rise*, UCAR Magazine, Fall, 2009.

(15) Church, J. A., *Estimate of the regional distribution of sea level rise over the 1950-2000 period*, J. of Climate, 17, 2004, p, 2609-2625.

(16) Helvarg, David, *Fiddling While Antarctica Burns*, New York Times, March 7, 1999.

(17) *Antarctica Global Warming - Easy*, Cool Antarctica, undated. See: http://www.coolantarctica.com/schools/Easy/global_warming_easy.htm.

(18) Stott, Philip, *Cold Comfort for 'Global Warming'*, New York Times, March 25, 2002. Stott is the emeritus professor of biogeography at London University.

(19a) Bentley, C. R., *Rapid sea-level rise from a West Antarctic ice-sheet collapse: a short term perspective*, J. of Glaciology, Vol. 44, no. 146, pp 157-163, 1998, as posted on CSA Illumina.

19b) *ACTIVE-RECENT SUBGLACIAL VOLCANISM WITHIN THE WEST ANTARCTIC RIFT SYSTEM* - - 2010 GSA Denver Annual Meeting (31 October – 3 November 2010)

19c) Barnes, E., *As Delegates Meet inCancun, Critics Say UN is Wrong Venue for Climate Change Debate*, FoxNews.com, December 2, 2010.

(20) Environmental News Network staff, *West Antarctic Ice Sheet not in jeopardy*, CNN Interactive, December 1, 1998. See Internet, www.cnn.com/TECH/ science/9812/01/

(21) Leake, J. and Hasting, C., *World misled over Himalayan meltdown*, The Sunday Times, January 17, 2010, as posted on Timesonline.

(22) Harvey, Claude, *2009 paper confirming IPCC sea level conclusions withdrawn, mistakes cited*, See: Watts Up With That? web site, February 21, 2010.

(23) IPCC, *Climate Change 2007 - The Science of Climate Change*, Cambridge U. Press, (1996).

(24) Sidell, M. Et al, *Retraction: Constraints on future sea-level rise from past sea-level changes*, Nature Geoscience, July 26, 2009.

(25) Verneer, M. And Rahmstorf, S., *Global sea level linked to global temperature*, Proceedings of the National Academy of Sciences Early Edition, December 4, 2009.

(26) Adam, David, *Climate scientists withdraw journal claims of rising sea levels,* gaurdian.co.uk, February 21, 2010.

5. All About California

5.1 Just Look What They Have Done to Energy: a Harbinger of Our Future.

• **California Situation - 2001.**

In my first book[1.1], the energy problems and political trends in California were reviewed. The conclusion was made that the far-lefters have created havoc in many areas of California. Indeed, charges[2] were noted that California has been turned into a socialist state. California was a perfect setting for the far-lefters. And their success was almost complete. The emergence of a new governor, Arnold Alois Schwarzenegger (AAS), offered hope that these far-lefters could still be contained. However AAS appears to have turned very green over the time spent in office.

• **California Situation - 2009.**

In an essay for Time[1.2], the energy dreams and trends in California are reviewed. The conclusion was made that "when it comes to energy California is not just ahead of the game; it is playing a different game." The article asks: "The End of California?" This is followed by a picture of AAS with the headline: "Dream On!" It is perhaps too soon to decide which of these two futures will win out. However there is little question that this state leads the way in going green and in embracing alternative energy.

Now one should not be surprised at this situation, Of all the states California has more weather related phenomena to be hyped due to the possible impacts of global warming. Some concerns:
· too little rain;
· too much rain;
· mud slides;
· too little snow-pack;
· too much snow-pack;
· grass fires;
· forest fires;
· Santa Ana winds;
· El Nino/La Nina and
· sea level rise.

Clearly each of these areas would need to be examined in detail to see what, if any, real concerns exist due to any global warming. And don't forget there is always the opportunity for hype such as the inclusion of pictures of San Francisco Bay in Al Gore's book AIT.

• California NIMBYism.

California has seen many serious problems over it's history. Five years ago their energy crisis just might be the biggest one the state had ever faced, at least up to that time. The effects of many years of mind-boggling bumbling—by consumer, energy and environmental activists; by many state politicians, and state energy bureaucrats; and by some who populate the major energy vendors and state utilities—started to become visible around 2000. California became the most difficult place in the world to build a major power plant. NIMBYism—the Not In My Back Yard syndrome—had taken hold.

NIMBYism rapidly grew into an epidemic in California. This diseased attitude not only focused on nuclear plants, but also on coal fired stations, electric transmission lines, off-shore oil wells, and natural gas pipelines. What has been built in California instead is thousands of Alternative Energy (AE) units[3].

• California Electrical Crisis.

While there were some harbingers of the power crisis in 1998 and 1999, it finally hit in 2000. A serious and steady price escalation started in April. Wholesale rates doubled, then doubled again in the third quarter and almost doubled again by the end of the year. Periodic and serious price spikes occurred several times, sending the California market into panic. Next the state was hit by a series of rolling blackouts. Needless to say the credit worthiness of the state electric utilities was destroyed, One utility filed for bankruptcy. Next the state started to buy power for the utilities, borrowing billions from the general fund; arranging for bridge loans of $4 billion, then another $5 billion; and ultimately arranging to issue new bonds on the order of $13 billion. "As a result of this energy crisis the bonded indebtedness of the state, will grow by 80% to deal with a problem that didn't exist even a year ago and would not exist now were it not for the early political decision against rate increases[4]." The state now not only has an energy crisis, but also a financial crisis, with the credit worthiness of the state starting to crumble.

Many now believe that deregulation of the electric industry was the cause for this crisis. However, as noted in my first book, several causes existed. These included:

· an incredible embrace of the Nimbyism attitude;
· alternative energy;
· the Federal government;
· the deregulation bandwaagon; and
· the deregulation plan.

Many view energy deregulation as a massive failure. What they ignore is the California plan just might be the dumbest deregulation plan ever conceived. The founders of this great state must all be turning over in their graves at the humiliation their state is now going through.

While there is no argument with the proposition that AE can be very useful, and save energy resources, it is the total reliance on AE that is one of the major reasons for the current predicament. However, the resultant contribution is small, and additional power generation capacity is still needed to back up this source, or the over all reliability of the system would be compromised.

And this is what California has built by embracing NIMBYism and AE: a compromised electrical power system. In industry, one of the key responsibilities of the head energy executive is supply security. Clearly there is no energy supply security in California today.

The California leftists completely ignored warnings that their system design was seriously inadequate at best and possibly unstable at worst. One report[5] on this subject listed a half a dozen major warnings from knowledgeable observers. And this is what California has built by becoming blinded by the deregulation potential, and ignoring red flags that their plan was seriously flawed: a vulnerable electrical system.

On March 31st, 1998 California's new *deregulated market* opened for business. The system worked superbly, and ISO officials celebrated with a lightbulb shaped cake. Brochures were available at this party that were entitled "Securing Reliability[5]." It would be three months before the party ended. The history of price behavior oer the net couple of years should have been a warning as to what can happen to energy prices under conditions of major stress on thee system.

On July 9, 1998, the price for reserve power was running at around $1/mwh. Then suddenly it spiked up to $5000/mwh, where it stayed there for three hours! Then it dropped back to the $1/mwh level. Four days later it shot up to $9999/mwh for four hours. One executive, that had earlier warned state officials, said these spikes were signs of someone probing the system, looking for the weak link.

Over 1999 wholesale prices were fairly stable. However, they did escalate from $20/mwh in the first quarter to double that rate by the fourth quarter. And for a few days in October and November prices exceeded $70/mwh.

For the first quarter in 2000 prices were back down to about $30/mwh. However, prices moved up dramatically over the next three quarters: $70, $135, and finally $230/mwh. In particular, over the second half of the year, daily average prices exceeded $150/mwh for 38 days. For four of these days, prices exceeded $600/mwh.

Several[6] recommendations were noted in my first book:
· don't make quick and irreversible commitments such as trying to isolate California from the western electrical market;
· don't turn the state into the permanent purchasing authority;
· don't spend taxpayer funds on large energy related projects;
· don't over commit on long term power contracts; and
· don't nationalize the state's electrical system.

In industry, one of the key responsibilities of the head energy executive is price security. Clearly there was no energy price security in California, and the sixth largest economy in the world is now, essentially, almost, a socialistic government. If this pathway continues the socialists victory in California will be rather pyrrhic, for they undoubtedly will lead California down the pathway to becoming a *third world nation*.

Yes, the socialists have almost succeeded in California. But perhaps they have over-stepped in their effort to create a socialistic energy dreamland. The crisis in California "is the direct result of leaders and citizens who believe they can legislate and regulate economic reality out of existence and who appear ready to stick to their fantasy until the lights go out and the economy crumbles[2]. This is what California has built by embracing NIMBYism, AE, socialism, and an unshakable conviction in California's ability to finesse every situation: an energy system that is severely compromised, with no supply security; an energy system that is highly unstable with no price security and finally a financial system that is surely tarnished, including a major reduction in the overall credit worthiness of the state. Yet in spite of this status CA politicians can still support projects such as a $578 million K-12 school.

• **United States Energy.**

Today, our current administration—by embracing NIMBYism, AE, socialism, and an unshakable conviction in their ability to finesse every situation—is bent on totaling remaking America. The majority of tax payers and key influences in California were mesmerized and brain washed and walked, like lemmings, over the cliff. Is this not what is going on in the U. S. today? Gail Heriot, a Professor of Law at the University of San Diego, and a member of the US Civil Rights Commission urges caution[9] over "the dangers of precipitous action especially when its advocates appear to be caught up in something akin to religious fervor." She noted that her "instincts run toward stop, take a deep breath, and be absolutely sure that you're not about to put the world's economy in a stranglehold just to please the people who despise modernity." She reported on an incredible con job that was unfamiliar to this writer. See NOTE 10 for more details.

Today we have also moved in to the era of massive bailouts: banks, insurance companies, the auto giants, and, yes, the state giants. And California—what with a staggering $41B deficit by mid 2010—leads the way[7] with an estimated $26B from the new stimulus package.

A major fraction of this package will go to wind energy. Today, we see rather incredible support for massive wind energy projects. One writer[10] scolds the Sierra Club for its general support for wind energy. It asks how can such a looming industrial presence as wind farms "be reconciled with the goals of maintaining choice natural habitat, while reducing the impact of human activity." This writer reported on a 2000 wind turbine project, along the Allegheny Mountain ridges, that would impact 40,000 acres. He argues that support for such projects "betray sound environmental and scientific precepts, ideas that many knowledgeable environmentalists hold dear, while putting at risk vulnerable species and valuable habitat - - -."

The above paragraph could have stated that a major fraction of this package will actually go to not just wind, but to all AE. However, the conclusion would be the same. Now if one would include nuclear as a type of AE there might be some rational. However, as one who spent a tiny part of his career on trying to get a Michigan nuclear project rolling—about half a dozen tiny assignments, only to see it ultimately converted to a natural gas plant— I am well aware of difficulty to get nuclear plants built.

This nuclear project triggered dozens of letters, perhaps over a hundred letters, to the local newspaper - The Midland Daily News (MDN). I may have contributed over a dozen myself, as I debated the projects merits with a Mrs. Mary Sinclair, the local anti-nuclear activist and her acolytes. Several of these letters are highlighted in NOTE 11 .

With all due humility, I can claim I won this battle—Mrs. Sinclair was not elected. However, she can claim she won the war, as the two nuclear reactors were never completed. One was about 95% done and the other about 80% before the utility called it quits.

A more detailed analysis of our national energy situation, covering all energy options is covered below in §6.

• Cap and Trade Initiatives.

(1) <u>Definition</u>. Very briefly a Cap and Trade (C&T) system, in a pollution context, is one where a polluter is given a target level of pollution. For example if you are allowed to emit 100 tons, say of carbon dioxiide, over a certain interval, but you only emit 80 you have 20 credits which can be sold. However, if you emit 125 tons you would need to buy 25 credits at the prevailing price on the "carbon exchange."

(2) <u>California Involvement</u>. AAS signed [8] a C&T bill in 2006. However, this bill, if implemented fully, will send CA back to the days of "The Flintstones." It calls for reduced GHG emissions to 1990 levels by 2020, followed by an 80% reduction below the 1990 levels, by 2050.

Today, CA is in the spotlight as a possible model for the national system. With Henry Waxman now in charge of the Energy & Commerce Committee, the odds are high that the CA laws will be the model for the national system. To meet the above goals a C&T program must be designed and adopted by January 1, 2011, and the program must begin in 2012. What is not reported in these and other references on C&T is where the implementation of this CA law will take the state. Can the population of CA truly reduce carbon emissions even to their 1990 levels, let alone 50% below that? I think not. The odds are this will push CA further towards becoming the numero uno ward of the Federalies.

This law also calls for CA utilities to get a third of their power from renewable resources by 2020 with 20 percent by 2010. Renewable resources includes the usual suspects plus "small hydro" projects. Note that excludes "large hydro", the kind of hydro that has helped make CA great. However, the utilities[11] are "woefully behind" in meeting these deadlines. As of 2008:

· Southern California Edison had achieved 15.5 percent;
· Pacific Gas and Electric was at 11.9 percent and;
· San Diego Gas and Electric only at 6.1 percent.

These three utilities will not meet 20 percent until 2013, at least.

Now one of the so-called positives for this renewable resource initiative is its impact on jobs. However, none other than William Jefferson Clinton has recognized[12] that such energy utilization has actually costs jobs in Spain. His source for this position was a study done by Professor Gabriel Calzada. This report "argues that every job in renewable energies created in Spain in the year 2000 has - - - been the cause of the loss of 2.2 jobs elsewhere ion the economy.

A few days later the following statement[13] was released: "efforts continue to be made in the U. S to discredit the Spanish green jobs study, and even personally attack its author - - -." This statement went on: "More than 10 years and nearly $40 billion in public investment later, Spain still only acquires less than one percent of its power from solar, and the vast majority of so-called green jobs created by the government to support that industry are no longer in existence today." One can't help wonder if this is really the pathway to "near-term economic growth and-long term energy security."

The thoughts[14] of George Will on this subject included that one could be: "dismayed by the frequency with which such findings are ignored simply because they question policies that are so invested with righteousness that methodical economic reasoning about their costs and benefits seems unimportant. "When the president speaks of 'new green energy economics' creating 'countless well-paying jobs,' perhaps they really are countless, meaning being incapable of being counted."

However, Californians have ascended to positions of power in Congress and in the Obama administration. And many of them were heavily involved in writing the $800B stimulus bill that was passed in to law in February, 2009. However, it is possible that the opposition to the C&T law maybe at an all time high

(3) <u>The House Bill.</u> "Mr. Waxman (for himself and Mr. Markey of Massachusetts) introduced" the W-M bill. This bill[15] has some rather modest charges, such as: "to create energy jobs, achieve energy independence, reduce global warming pollution and transition to a clean energy economy." It passed by a 219-212 margin, with 44 Democrats defecting and only eight Republicans joining the majority.

Now out of the many steps defined in this bill one of the more outlandish is the call for the creation of "Clean Energy Innovation Centers." How may? Where? At what cost? In any event these centers are charged with annual savings as shown below.

Year	Oil M barrels	Oil %	Gas BCF	Gas %	Electricity Bkwhs	Electricity %
2020	22	0.3	116	0.5	14.4	0.13
2p3p	111	1.6	2018	8.7	167.0	1.5

Energy Innovation Centers: Annual Absolute and Percent Energy Savings

While the percent savings are small, the achievement of the absolute amounts would represent very difficult goals. There is no low hanging fruit out there. It is doubtful whoever wrote such inputs has had any experience in product development or technology development. The time elements in technology and product development can be absolutely relentless. I have noted in §6.4—specifically for the Sodium Sulfur battery and for the Magnesium dry cell—just how relentless this can be. The reader is also referred to Reference 11.3 for that section.

Job creation is another claim for this bill. The Spanish experience noted above is surely food for thought on that claim.

There is no question this bill would create new jobs[16] for many new government bureaucrats and regulators. Examples:
· new real estate appraisal processes will see new training mandates for appraisers;
· each industry will have to calculate their its 'trade intensity' and 'greenhouse gas intensity' All material entering or leaving the country will interact with bureaucrats "to ensure the equitable international balance of carbon leakage", whatever that might be.

In conclusion it is hard to see the House Bill achieving any of the claims made for it. However, it is easy to see it as a huge, but hidden tax increase.

(4) The Senate Bill. Although this bill has been shepherded by senators John Kerry of Massachusetts and Joe Lieberman of Connecticut, there is still major western emphasis involved, such as Senator Barbara Boxer of California and Harry Reid[17] of Nevada. Now instead of "introducing this bill and sending it to committee"- - - "Reid will be taking the bill drafting process behind his closed doors for the deal making." Liebermans "cooperation with this secrecy effort is a good indication of why the country would be better off if he stayed in his own (Democratic) party and was defeated by a Republican" in 2012.

Now there are reports[18] that the Democrats will try to use the "reconciliation" process which only requires 51 votes to pass. But Senator Inhofe of Oklahoma reports, by his count, he "thinks they have just barely half that number, maybe 26 votes." He goes on: "the senate has voted on it so many times now, and each time they lose more and more support because people are aware that this is just another huge tax increase."

One of the reasons he is so confident is that the Democrats, in the past, when they needed a few Republican votes, would agree to loosen restrictions on offshore drilling. That trick would surely not be applicable today, after the Deepwater-Horizon drilling rig tragedy.

However, there is surprising corporate support[19] for Obama's Cap & Trade Legislation. The C3: Global Warming: Jobs/Profits/Taxes/Reparations blog provides a listing of 35 large corporate supporters. However five of these have given up on this support.

References and Notes.

(1.1) See reference (1) in the Preface.

(1.2) Grunwald, Michael, *The End of California? Dream On!*, Time, November 3, 2009.

(2) For example of a report on socialism in California see the work by "A Conservative News Forum", *Let's Laugh at California - They Deserve It*, March 27 2001. This report noted that "California is - - a failure of socialism. It is the direct result of leaders and citizens who believe they can legislate and regulate economic reality out of existence and who appear ready to stick to their fantasy until the lights go out and the economy crumbles". This report further noted "by regulating through ideology rather than according to economic reality the government of California has effectively bankrupted the utilities and is in the process of seizing their assets. This will change the role of government from that of regulator to that of owner and operator of the states electrical grid - The People's power company. How Soviet". Sounds incredibly familiar as to what is now going on in the U.S., but this was written eight years ago. Unfortunately, the laugh is on the U.S. taxpayer.

See: www.freerepublic.com/forum/a3ac14c4b2f25.htm.

(3) AE - Alternative Energy systems in California include geothermal, photovoltaic, other solar, waste energy, wind turbines and bio mass fired power plants.

(4) Flannigan, J., *Californian's Will Be Paying Energy Crisis Bill for Years*.

See: www.latimes,com/business/reports/power/lat_costs010329.htm, March 29, 2001.

(5) Stanton, S., *Special Report: How Californians got burned - - The state electrical system is in a shambles, and the worst may be ahead. How did things get to this point?*, The Sacramento Bee News, May 6, 2001.

(6) *Manifesto on the California Electricity Crisis*, Convened under the auspices of the Institute of Management, Innovation and Organization, University of California, 1-26-01. See: www.haas.berkeley.edu/news/california_electricity_crisis.html.

(7) Editorial: *Backward California*, Oil & Gas Journal, February 23, 2009.

(8) Editorial: *Cap and Trade Legislation: Will California's AB32 Go National?*, See www.triplpundit.com, February9, 2009.

(9) Heroit, Gail, Is Al Gore the Re-incarnation of the Xhosa Prophetess Nongqawuse, guest blog on Master Resource, July 11, 2009.

(10) Boone, Joe, *The Sierra Club: How Support for Industrial Wind Technology Subverts Its History, Betrays Its Mission, and Erodes Commitment to the Scientific Method (Part II)*, Master Resource, April 18, 2010.

(11) Lifsher, Marc, *Key points still debated regarding California renewable energy goals*, Los Angeles Times, August 29, 2009.

(12) Gallego, J., Cabellero, C., *Clinton: Green Energy "Has Cost Many Jobs,"* El Mundo, pg48, May 23, 2009.

(13) Henderson, Laura, Tucker, Chris, *Former president channels Prof. Gabriel Cazada in delivering veiled rebuke of Obama's Spanish-inspired green jobs plan*, posted on The Institute for Energy Research, May 27, 2009.

(14) Will, George F., A Quixotic Pursuit: Green Energy Jobs, The Washington Post, June 25, 2009.

(15) Bill, 11th Congress, 1st Session, H. R. 2454, May 15, 2009.

(16) *Required Reading: "What was in the Waxman-Markey 'Manager's Amendent'?"*, See Ironic Surrealism blogivists.com, June, 2009.

(17) Dunetz, Jeff, *Senate Cap and Trade Skipping Committee Process - Will Be Drafted Behind Closed Doors*, RedSate.com: Conservative Blog and News, April 15, 2010.

(18) Meyers, Jim, *Inhofe: Cap-and-Trade Has No Chance in Senate*, Newsmax.com, April 25, 2010.

(19) **C**$_3$ See http://wwwc3headlines.com/global-warming-economicsprofiteeringreparations/. The input of interest is the 8th one on this site, March 12, 2010.

5.2 A Salute to a California Pioneer

• **Background**

Late in 2008 I obtained word that I had received a very kind tribute to a letter[1] I had written on energy planning and global warming. The tribute was from a world famous environmentalist from California. Now skeptics—on the AGW issue, perhaps on any issue—seldom, if ever, receive accolades from the community to which they are skeptical. This is particularly true for the environmental community. Skeptics receive many insults and innuendos, but accolades, well hardly ever. Indeed, this was the first environmentalist accolade for me, so it was very special to me.

This environmentalist was Dr. Donald Anthrop, professor emeritus of environmental studies at San Jose State University. He has written many articles and letters not only on the environmental area, but also on various aspects of the energy spectrum. More on this *distinguished veteran* shortly.

•The Letter

The letter on energy and global warming asked if we really need to do such planning in a fossil-fuel constrained world. We still have a serious energy problem and many organizations are searching for the right energy policy, with most doing this in a fossil-fuel constrained world. However, this may be a terrible constraint as it is just possible that the diagnosis on anthropogenic global warming (AGW) is erroneous. Today citizens should demand an open debate on this fossil-fuel constraint, something that is not happening.

This letter next asked: What do the Sloan Professor of Meteorology at MIT, the founder of the Goddard Institute for Space Studies and the head of the space research laboratory in Saint Petersburg, Russia have in common? First, they are skeptics of the AGW hypothesis. They are not fans of Gore's *An Inconvenient Truth,* a part of the *political science* of alarmism that began 40 years ago, with Paul Ehrlich's *The Population Bomb.* And they are all physicists, namely: Richard Lindzen, Robert Jastrow and Habibullo Abdussamatov. Physicists just might hold the key to this diagnosis. This commentary is based on a paper[2] entitled *Global Warming — Witnesses for the Defense of the Skepical Perspective: Physicists.*

This letter then asked: how can a barely discernable 0.6°C increase since the late 19th century gain acceptance as the source of weather catastrophes? Answer: those with a vested interest in climate alarmism have cranked up the propaganda.

This interest in physicists started with a 2006 Houston Chronicle interview with Dr. Rapley of the British Antarctic Service. He asked: "if carbon is increasing, how can you deny there's going to be warming?" Rapley stated if you knew how physics worked, you would *stop arguing on* AGW.

The letter next asked: How then do physicists see AGW? As openers, a 1990 analysis by **Dr.** Jastrow, plus Drs. William Nierenberg and Frederic Seitz, (all deceased) gave a range of 0.4 — 1.8°C for the next century. This was much lower than the UN, as the Intergovernmental Panel on Climate Change (IPCC) reported in 1990, a range of 1.5 — 4.5°.C. The above physicists saw this as alarmist. While they agree, <u>if the assumptions they used are valid</u>, there is going to be a warming, but nowhere as big as the IPCC would like the public to believe. Then the IPCC report in 2001 boosted the range to 1.4 – 5.8ºC.

Dr. Lindzen, the MIT professor, noted in 1993, that model predictions depend on large increases in CO_2, plus mechanisms within the models that <u>amplify</u> the climatic response to increasing CO_2. These amplifications need to be debated. In 2006 Lindzen also noted that alarmists intimidate dissenting scientists. He noted that scientists who deviate from alarmism have seen their funds disappear.

Dr. Abdussamatov reported that Mars also has global warming. But parallel warmings on Mars and Earth can only be a result of an increase in the one factor common to both planets: solar irradiation. He now believes this has peaked and sees deep cooling by 2040.

Two additional witnesses, of the 17 covered in the energy economics paper, were also noted.

Dr. Sherwood Idso, a former research physicist at the U. S. Water Conservation Lab in Phoenix and an adjunct professor of Botany at Arizona State University has argued that increasing CO_2 levels are beneficial due to an increase in photosynthesis, leading to significant increases in crop and forest growth.

Dr. Hendrick Tennekes, former director of research, Royal Dutch Meteorological Institute, lost his post due to his views on climate modeling. He is concerned with the monopoly that modeling has on climate research. An example of the many short-comings with models: they don't include feedbacks between changing farming and forest practices and atmospheric circulation. For this and other reasons they can't agree on precipitation patterns, a far more relevant factor to food production than a tiny increase in temperature. He concluded: "We only understand ten percent of the climate issue."

Finally this letter drew several conclusions:
· the claim that we face an imminent catastrophe is unfounded and terribly inappropriate;
· the views of 17 physicists/mathematicians are proof of a serious debate on AGW;
· the claim that "all scientists agree" is juvenile at best, fraudulent at worst and
· all scientists need to reconsider the position of Thomas Huxley: skepticism is the highest of duties for scientists, blind faith the one unpardonable sin."

Indeed, with all of this testimony against the AGW hypothesis and climate alarmism, the letter asked one final question: is it not time to have this issue tossed out of court?

• **The Tribute**.

Anthrop wrote the editor of the journal[1] and extended accolades to myself for what he described as an excellent letter on global warming, and to the editor for publishing it. He noted he was particularly heartened by the writers quote from Thomas Huxley: "Skepticism is the highest of duties for scientists, blind faith the one unpardonable sin." Anthrop then observed that, unfortunately, skepticism is sorely lacking in academic circles, especially environmental programs.

• **The Environmentalist - Donald Anthrop**

Areas of interest, research and activity include:
· timber resources, California Redwoods;
· biofuels[3] in general, fallacy of ethanol policy - noted that 100% of the U.S. corn crop to ethanol would meet only 11.5% of gasoline demand;
· rivers, water quality, water resources;
· Federal and State water projects, water reclamation;
· agricultural drainage, erosion and sedimentation;
· evaporation ponds;
· birds in general with special attention to owls and hawks, waterfowl research;
· wildlife habitation;
· noise pollution.

• **The Energy Analyst - Donald Anthrop**

In addition to his classical environmental work, Anthrop has written extensively on energy. He has commented on many issues. A few examples, with very brief comments:
· fuel cell vehicles - sees them as a myth;
· energy independence - it wont happen, but high chemical costs, such as for ammonia, hurts;
· Hydrogen[4] - calls it an empty environmental promise;
· Kyoto Treaty[5] - sees this as a terribly flawed treaty, which will do serious harm to the economy;
· electric vehicles[6] and carbon emissions.

- **The Photographer - Donald Anthrop**

In addition to his efforts in the energy and environmental areas Donald Anthrop is a long-term photographer, who knew Ansel Adams personally. In 2007 he donated his superb collection of California landscapes and scenery to the American Lung Association.

- **Conclusions.**

His early work on the northern California redwood forest contributed to the establishment of the Redwood National Park in 1968.

Earlier in this book I introduced the concept of *Distinguished Veterans (DVs)* in the scientific and research fields. I noted four such DVs in the hurricane field. Since then I have written on the DVs in physics[15] and on agriculture, botany and food production. I have not written a paper yet on the DVs in the environmental field, but when I do Dr. Anthrop would head the field.

References and Notes

(1) Westbrook, Gerald T. *Warming Debate Needed*, Oil & Gas Journal, November 3, 2008.

(2) Westbrook, Gerald T. in the International Association for Energy Economics Energy Forum, Third Quarter, 2008.

(3) Anthrop, Donald F., *Biomass potential*, Oil & Gas Journal, September 5, 2005.

(4) Anthrop, Donald, *Hydrogen's Empty Environmental Promise*, Cato Institute, Paper No. 90, December 7, 2004.

(5) Anthrop, Donald, *The U.S. Carbon Emissions, and the Kyoto Protocol*, a point and counterpoint presentation, on climate change with Anthrop taking the counterpoint, Pacifica, Spring 2006.

(6) Anthrop, Donald, *Electric Vehicles and Carbon Emissions*, Contra Costa Times, May 22, 2010

6. The "Flight of the Phoenix" Revisited
We can live with a Fossil Fuel Future: Oil, Gas, Coal and Shale Oil

6.1 The Analogy.

The title of this essay[1] is *The "Flight of the Phoenix" Revisited*. This title is taken from a 1965 movie. For those who have not seen this rather superb effort, it featured a crash landing of a C-82 in the Sahara Desert in the 1950s. This was caused by a severe dust storm that drove their plane far off course. After a few days of waiting, the 11 survivors came to the conclusion that they were not going to be found by an air search.

The predicament of the survivors serves as an analogy to our energy situation. Both situations could be described as rather desperate. Both situations would exhibit deeply entrenched opinions, zero tolerance for other views, little real information and much confusion. Three options emerged that were equally unattractive, widely disparate, dangerous, and with low odds of success. They were: walk out; ride out in a camel caravan, if one should ever pass; or fly out in a re-created plane.

Two of the survivors died in an attempt to walk out, and two were killed by members of a small caravan. It finally became clear to the remaining seven that their only chance was to rebuild the plane. It was not until all other options were proved futile that the remaining survivors—very gradually, and very reluctantly—were able to coalesce around the one option that would save them, namely to re-build a plane from the wreckage.

Initially most saw this as an insane idea. However, the C-82, sometimes referred to as the *Sky Truck*, consisted of two long booms, each with a large engine, bracketing a large central hull. It was owned by an oil company, and its hull was full of oil field equipment: cutting torches, welders, winches, cable and so forth. So the opportunity to rebuild a plane was there, by salvaging the undamaged boom. And much like the legend, *The Phoenix* did "rise from the dead" and fly again, based on the same technology as the initial plane, but scaled down, and seven lives were saved.

Our challenge is similar, to re-build our energy system, but one that still relies on fossil fuels. For the *Phoenix* survivors, the critical resource was water, with about 12 days of supply left to complete their reconstruction. For our country, the critical resource is oil. And just as the *Phoenix* survivors rebuilt their plane, our challenge will be to rebuild our energy system, albeit with still major reliance on oil.

Item	Phoenix Survivors	U. S. Economy
Original System	Two engine plane	Based on oil
Reconstructed System	One engine plane	Based on oil, but lower share

Comparison of Major Reconstruction Efforts.

Whatever the ultimate mix of energy resources turns out to be, this re-creation will be very difficult, and will take time -- and at least some failure of the other options -- before the activists get on board. One can only hope our country will exhibit the same ingenuity, tenacity and success in solving our energy crisis that the survivors of the *Phoenix* did.

6.2 On Our Energy Situation, Price Stability and Supply Security

When the price of crude oil sky-rocketed from $95 to $147 per barrel, it was easy to see the "clear and present" danger of such a situation. Less clear is the impact with a price retrenchment from $147 to $35 per barrel. In such a situation energy planning and implementation becomes almost impossible. In particular coal gasification, shale oil processing and many alternative energy (AE) ideas become almost impossible to fund.

Today, with the trauma in our financial system, and the slippage in oil prices, it is hard to spend much time on long term energy planning. However, our country is most likely still in a serious energy situation. Perhaps no one has stated it stronger, and perhaps more pessimistically. than Matthew R. Simmons, a Houston investment banker, in his many presentations[2.1] on the state of the oil and gas system. "It is sick", he concludes, and "its vital signs are terrible." Simmons asks many questions in his talks, including "does anyone know a price for oil?"

This theme was continued in a commentary[2.2] in The New York Times. In this commentary the writers note that "Compasses are spinning." They went on: "Everyone seems uncertain whether increase in supply or decrease in demand will affect prices as they have in the past." Indeed, "some wonder whether the market is broken in some way creating a bubble of artificially expensive oil."

While the price at $147 may now look like a bubble of artificially expensive oil, Simmons thinks not. He notes[2.3] that much of the global supply is from fields that are really old and getting older. He sees the idea that OPEC has major spare crude production capacity as a myth. And he concludes "the era of cheap oil is over." So it is highly likely we will see the era of expensive oil once again and soon. Now, not surprisingly, there are those who disagree with his rather pessimistic outlook. Amy Jaffe, director of the Energy Forum at the James A. Baker III Institute for Public Policy, is far more optimistic on both gas and oil. See her talk at USAEE Luncheon, November 18, 2010. Jaffe was asked by this writer if she was too optimistic on this area. She did not think so. She defended her gas forecast intelligently and vigorously.

One can see three main pathways into the future. These pathways— more precisely areas of concentration—might be viewed, by some, as equally unattractive, widely disparate, dangerous and with low odds of success. These areas of concentration are:
· alternative energies,
· fleet electrification/nuclear power, and
· fossil fuels.

The title for this essay states that we can live with a fossil fuel future. *Warmers* and environmentalists will see this as "an insane idea" Indeed, our new energy secretary, Steven Chu, has stated [3.1] that 'Coal is My Worst Nightmare." I would argue we don't have the luxury to "turn up our noses" on the use of this fuel.

• Price Stability.

I will further argue that a key pathway to price stability is to reduce our call on global oil. What is necessary is an aggressive move to reduce our call on global oil, not to eliminate it. Surely the USA is not the only major buyer of global oil. And China and India have become very big overnight, and will be very big in the future. But the USA is viewed as one that should and could take action.

Now, oil speculators are broadly accused of manipulating the market. Possibly, but all they do is simply buy and sell crude oil contracts. In order to perform this task, perhaps many times a day, they strive to read the market. Here I argue that the USA's call on global oil is an indicator to such risk takers. If the speculators of this world are convinced that we are not going to put our energy house in order, if they can't see any progress in our ability to cut demand, or in our ability to increase supplies, they will soon start to bet again that our call on oil will go up. And, in the absence of any other inputs, they will, once again, place their bets that future prices will be a bit higher and not lower. And their next bids will also be a bit higher.

I will argue that the best of the three non-perfect, and drastically different policy initiatives, is a fossil fuel focus. Some will think such a pathway is absolutely insane. However, the hope is that, in the future, they will come to the conclusion that it is the only way out of this situation. Here we would boost our conventional supplies of domestic oil and gas, we would increase our use of coal generation and we would start to utilize coal gasification, coal liquefaction and shale oil. While not pretty, this pathway has the best odds for reducing our call on global oil, and the best chance at providing large amounts of additional energy for the future.

I will argue that the other paths—a huge expansion of nuclear plus fleet electrification, and an intensive alternative energy focus—will not lead to the price stability we need.

• Supply Security.

This interest in supply security is in no way a pitch for energy independence. This in no way is another "Project Independence" Indeed, energy independence is seen as an impossibility and indeed, as an erroneous objective. Supply security can be achieved by building on a strong and reliable mix of all supply options. This includes all domestic options, including coal, offshore oil and Alaskan oil, specifically the Arctic National Wildlife Refuge (ANWR). Now the environmentalists insist we can't touch the ANWR oil - the area is so pristine. I would argue we don't have the luxury to "turn up our noses" on the use of this resource. Many presentations of this topic describe it as a pristine refuge, a cathedral of nature and America's Serengeti. They inevitably show a picture with magnificent mountains in the background - the Brooks Range. But these mountains are about 50 miles from the featureless coastal plain.

George Will[2.4] raises the possibility that "some people use environmental causes and rhetoric not to change the political climate for the purpose of environmental improvement. "Rather, for them, changing society is the end, and environmental policies are mere means to that end. "The unending argument in political philosophy concerns adjusting the balance between freedom and equality." The overall good "is to enlarge government supervision of individuals lives."

Supply security also includes oil from the Mid East and syn-crude from Canada. One might ask why does it make sense for multi billion dollar investments in Canada for tar sands mining, processing and upgrading, but not an analogous effort in the U.S. on coal and oil shale mining, processing and upgrading? How much difference can there be, when crude prices exceed $100 per barrel?

• **Our Coal Situation**.

The sub-title for this essay states that we can live with a fossil fuel future. *Warmers* and environmentalists will see this as "an insane idea." Indeed, our new energy secretary, Steven Chu, has stated [3.1] that "Coal is My Worst Nightmare." I would argue that we don't have the luxury to "turn up our noses" on the use of this fuel. I would say it is a very necessary resource.

One of the purposes of this essay is to put in a good word for coal. It would seem that coal has become the pariah of any energy plan. Even without the so-called global warming crisis, coal has been viewed by many activists as unacceptable, due to the criteria pollutants (sulfur oxides, nitrogen oxides, and particular matter) and mercury emissions that it emits. Many activists have tended to ignore that the criteria pollutants have been effectively controlled[3.2] at an affordable cost, and efforts are underway on mercury emissions.

Perhaps the activists have recognized for a long time that they needed something more—to shackle coal the same way they have shackled nuclear energy—and that is where the global warming issue comes in. Hence we have had a tidal wave of publications against coal. And the conventional wisdom has become that this threat of anthropogenic global warming (AGW) is worse than high level radiation from fuel rod disposal, and certainly worse than international terrorism.

Many critics of coal insist[3.3] we must move into a carbon constrained world, even though we are facing the most serious energy situation ever. Of course our politicians have jumped in, many with well meaning, but terribly misguided ideas, plans and bills. For example the Democrats have come up with such winners as suing[2.5] OPEC, or blame the Exxons of this world, and threaten the imposition of windfall profits taxes, and if that doesn't intimidate them into coming up with more oil and cutting prices, to nationalize the oil industry. And these critics have come up with the phrase: *clean coal*. Any coal that doesn't meet their definition of clean coal—which is coal used in a carbon constrained plant—is dirty or ugly or filthy coal. This definition needs to be challenged. Later in this essay I will go into my rather deep experience with coal and my reasons as to why I think this so-called clean-coal "edict" can be challenged.

• Political Positions on Climate Change.

The climate change situation facing this country has been described as the most awesome threat our country has ever faced. Barack Obama[4.1], as a senator, has stated "The future of our planet is at stake." Harry Reid called[4.2] "climate change the most important issue facing the world today." Well not hardly, as the senate shelved this awesome issue after only 3½ days of debate. And surely Nancy Pelosi cannot be ignored. Now as an aside I have commented in Section1 on the P, P and P test. Perhaps this test should be written as the Pelosi Test, but that can be left for later writing. Here we are interested in her views on climate. And Pelosi has commented[4.3] on climate change with her council that we must start cutting global warming pollution immediately, "to avoid catastrophic climate impacts."

These, and other politicians, seem to have two naive convictions and/ or political positions.

(1) **All climate alarmism is true**. Many politicians seem to have zero comprehension of the huge uncertainties wrapped up in the AGW issue and the mantra of fear utilized by its supporters. They accept as gospel that society must move in to a carbon constrained world.

(2) **All Alternative energy hype is true**. They also accept, with blind faith, that society will find alternative energies (AEs) rapidly and in huge quantities so we can replace oil, gas and coal painlessly. All we need to do is pour money into research and bet on incredible, major, long shots.

6.3 Emerging Strategies - I: Focus on Alternative Energies.

The *warmers* and activists argue that we so have many types of AE from which to choose: solar, wind, hydrogen, etc, etc, etc. However, they seem to have no comprehension of the odds of success, nor of the science, technology and investment required to make these AEs economical. Several types of AE are reviewed in this section.

Wind Energy.

(1) <u>Recent growth</u>. In previous writings I noted that although California had built over 15,000 windmills by 2000, these produced power only 18% of the time and contributed only 1¼% of the total state generation. Since that time, growth of this energy source in Texas has been notable. In 2006, for example, Texas passed California as the state with the most wind capacity as noted below.

Year	Texas - mws	California - mws	U. S. - mws
2000	181	1,646	2,566
2005	1,995	2,150	9,149
2006	2,739	2,376	11,575
2007	4,296	2,450	16,596

Key Installed Wind Capacity - mws

(2) <u>Key problems with wind energy</u>. Surely, much of the above growth in Texas is due to government support at all levels. In spite of this support, however, there are many problems.

(i) <u>High equipment and installation costs</u>, even with many subsidies.

(ii) <u>Limited availability of the installed capacity</u> and, hence, limited generated power.

(iii) <u>Very remote locations</u> frequently requiring new and long transmission lines.

(iv) <u>Impact of working with governments</u>. While government support may help get a market started, it comes at a price. For an example, the reader is referred to a recent *Houston Chronicle* commentary[5.1] on what government involvement can lead to, when it gets into market creation for an AE such as wind energy. The writer was reporting on a conference in Houston on wind energy that was attended by 10,000 people. He commented that the more the hype blows, the more Robert Bradley -- a former director of public policy analysis for Enron -- "hears the voice of his old boss, Ken Lay." Bradley spent 16 years at Enron, and the writer noted that his ex company "was a major backer of the state's 1999 mandate calling for the development of renewable energy sources, including wind generation." So he asked: "Is it really the winds of change we're hearing? ... or just Enron's ghost laughing?"

A more detailed essay[5.2] by Bradley can be found in a brochure from Lindenwood University, wherein he noted that "Enron lived, thrived, and perished in and through the mixed economy. Enron's artificial boom and decisive bust had more to do with government regulation than free markets. Ken Lay's meteoric rise and stunning fall were not the saga of a capitalist wildcatter, they were the tragedy of a political rent-seeker in action, prominently including government intervention sought in the pretext of addressing climate change and promoting 'green' energy."

For more on rent-seekers, wind energy, Enron and Ken Lay, the reader is referred to a book[5.3] by C. Horner, especially the section entitled *Enron: Leader of the Axis of E's*. Here one will find dozens of nuggets, such as President Clinton naming Lay "to his exclusive panel of insiders, the Council of Sustainable Development." This administration liked Lay for a variety of reasons, including that he would represent "an icon of responsibility in their GW endeavor."

(v) <u>System penetration</u>. This attribute was described in a brief summary[5.4] of the hype used by politicians, plus some key comments on system integration. The case study is from the Canadian province of New Brunswick (NB). The energy minister for NB started with a claim that they would add 4,500 mws of wind energy to an existing system of 4,000 mws. He made this claim when the neighboring state of Maine, which had twice the population of NB, had all of 42 mws installed. He then back tracked to 1,250 to 2,000 mws. Setting aside how many turbines they might actually want, there is the question of system design. Even adding as little as 15 percent wind generation to an existing system requires constant monitoring and adjustment to prevent power fluctuations and grid instability. However, the writer cites German input that this should be more like four percent or 160 mw. He concludes that obtaining 1,250 mws from wind in NB is unrealistic, and that the idea of 4,500 mws is an "impossible dream."

(vi) <u>Wind turbine syndrome</u>. A Dr. Robert McMurray, a former dean of medicine at the University of Western Ontario, has reported[5.5] on a recent survey he conducted of people living near wind turbines. He found the majority suffered from headaches, sleep disturbances and depression. There have been other surveys. Google lists 55,500 response to a search on this subject. Other medical issues have been noted. One can only file this problem as a work in progress, in this listing.

(vii) <u>Bird Kills</u>. A similar search on this topic showed 5,480 responses. While only a tenth of the above category, this problem has not gone away.

(3) <u>Conclusions on wind energy</u>. With this panorama of problems, it is hard to believe that this source of energy -- in spite of major government support and promoters such as T. Boone Pickens -- will be the solution to our crisis. Perhaps the prognosis that is the most interesting comes from a 2002 essay[5.6] on net energy. Two conclusions stand out:

(i) <u>Most optimistic assumptions</u>. Even under such conditions, this analysis suggests that "Wind Power is capable of furnishing only a fraction of the net energy needed to power the U.S."

(ii) <u>Machines constructed everywhere it is practicable</u>. However, even under such a scenario, "they would only be able to provide a pitiful amount of the net energy" needed.

The above very negative comments were published in a sustainable energy forum, a forum one would expect would be very supportive of wind energy. However, any conclusion on the future of wind energy should be put on hold until the future of high energy density batteries is known (see Section 4.2, below).

Biofuels.

There is little question that biofuels will play a role in our energy system. However, projections range from unbelievably bullish to an area beset by many problems. One input[6.1] comes from a key environmentalist. This scientist avoided the problems of the efficiencies of processes to produce ethanol and questions on *energy returns on energy input* (EROEI). Rather it asked the question on the amount of electrical generation that could be achieved from wood grown on the net forest area of commercial timberland, namely 483 million acres. The answer was 17.5 percent. He concluded that the idea that biofuels are going to end US dependence on foreign energy supplies "is an illusion." Another reference[6.2] reported world consumption of 10 billion gallons of biofuels in 2006, including ethanol and biodiesel, noting that "this is 650,000 barrels per day, but less than 0.8% of worldwide crude oil consumption."

A summary of the potential, and problems, of biofuels follows.

(1) <u>Basic Ethanol Production</u>.

(i) <u>Ethanol from corn</u>. While modestly boosting the supply of liquid fuels, this fuel may actually be increasing our overall energy demand.

Many references report that this fuel requires more energy to produce than it delivers. Bloomberg[7.1] reports that David Pimentel, a Cornell University of Ecology and Agriculture, argues it uses 29 percent more energy than it delivers. Warmers argue the opposite. For example consider a story[7.2] on "a fuel ethanol enthusiast" that reported elation on hearing that the EROEI "had been greatly increased from 1.38 to 2." When informed that, even if that increase was true, "that was not nearly high enough" he became incredulous, outraged and insulting. This writer compared this value for ethanol to a value of 10 for U. S. oil, in 2000, for a very mature oil industry.

Ethanol enthusiasts seem to prefer talking about the oil imports displaced. For example one reference[7.3] cited 170, or 206 or 500 million barrels of oil import reductions due to ethanol production in 2006. These comments were by the Renewable Fuels Association (RFA) or its friends. However, the RFA also noted that ethanol production amounted to 4.86 billion gallons for 2006 or 116 million barrels. However, since ethanol has about two-thirds[7.4] the energy content of gasoline, the116 million barrels drops to 77 million barrels or only 2¼ percent of U.S. gasoline consumption. There is little question this level will grow. Even without the recent election victory there were many signs of major new growth coming. And the farm lobby remains very strong. Today, this incredibly subsidized field—corn subsidies and ethanol subsidies—would make Ken Lay rather jealous. Bloomberg noted[7.1] that the federal government has 20 separate laws to boost ethanol use, and 49 states offer additional support.

However, the future for ethanol is not all rosy. There are problems. For example it is now rapidly impacting many food markets such as beef, butter, cheese and milk. It is also putting many businesses at risk, such as dairy farms and cattle ranches. Dr. Steven Chu, has been [6.3] a staunch opponent of ethanol from corn, as have others[7.6, 7.7]. However, in spite of such expressions of concern Barack Obama is from a corn state and is supportive of the farm lobby. Hence significantly larger mandates for ethanol use are very likely.

(ii) Ethanol from sugar cane. Supporters frequently point to Brazil and argue "if they can do it, so can we." However, they conveniently ignore that all of the ethanol produced in Brazil comes from a far superior feedstock - sugar cane, not corn. They also forget to mention that Brazil, in parallel with their ethanol efforts, has also become one of the giants in offshore drilling and production. While a major fuel in Brazil, albeit subsidized , the prospects for this fuel in the U. S. Have been rather dismal. First

Brazilian sugar cane production in 2007 was over 100 times greater, 528 million tons versus only 3.7million. And U. S. Production is located only in Florida and Louisiana, although other feedstocks such as sugar beets and molasses, could also be used. While there seems to be little chance that ethanol from sugar will become a significant fuel, this option might best be filed under work in progress..

(2) <u>Second generation biofuels</u>. Dr. Chu's key work at the Lawrence Berkeley National Laboatory (LBNL) was the so-called Helios Project. This included the Energy Sciences Institute, a joint effort by BP, Cal-Berkeley, Dupont, the University of Illinois and LBNL. Unlike his position on corn ethanol, Dr. Chu has been reported [6.3] to be a big proponent of cellulosic ethanol. Hence one can expect an acceleration of effort in this area.

(i) <u>Ethanol from cellulosic feedstocks.</u> A recent report[6.4] by Sandia Natinal Labs and General Motors claim that "the U. S. could produce enough ethanol to displace nearly a third of all gasolines use by 2030." The report cited "that annual ethanol production from plant waste and energy crops could reach 90 billion gallons by that date, with 75 billion from cellulosic feedstocks such as switch-grass, corn stover, wheat straw and woody crops."

The 90 and 75 billion gallons are 2140 and 1790 million barrels or 5.9 and 4.9 million barrels per day, which represent 63 and 53 percent of our gasoline consumption – even more bullish than the news item cited. This article did quote two scientists.

(a) Cole Gustafson, North Dakota state University biofuels economist - "the 90 billion figure is the most aggressive he's heard to date, far surpassing a federal mandate calling for 36 billion gallons of renewable fuel to be blended into gasoline by 2022." He went on. "I really question if we can even make that." "This technology has been very slow to evolve."

(b) David Pimentel, Cornell University - called the 90 billion gallon number "off the wall." A more reasonable number - "10 billion gallons a year from cellulosic ethanol within the next 10 years."

In spite of such concerns significantly larger efforts are very likely in this area.

(ii) <u>Biodiesel</u>. This field [8.1] has had "some of the highest growth rates ever seen in the chemical industry." For example global growth from 2002 to 2007 was over 50% per year. And for 2007 growth was at the triple digit level.

A wide variety of feedstocks have been used: vegetable oils; animal fats and oils; and other biological waste items such as cooking oils and grease.

This spectrum of feedstocks has led to plants all over the globe, ranging from local to industrial size. Key feeds include soybeans and palm oil.

Major potential has been identified for most countries, with Malaysia, Indonesia, Argentina, the U.S. and Brazil leading the way. Table 2 summarizes the top 15 countries, showing a potential of almost 800,000 barrels/day and accounting for nearly 90 percent of the global total.

Country	B liters Per year	K barrels Per day
Malaysia	14.54	251.6
Indonesia	7.61	131.4
Argentina	5.26	90.9
United States	3.21	55.6
Brazil	2.57	44.4
Netherlands	2.50	43.2
Germany	2.02	35.0
Totals	37.21	652.1
Report totals	51.00	882.3

Top Seven Potential Biodiesel Countries

(iii) <u>Butanol based</u>. The LBNL is one of several organizations persuing this chemical. In a recent speech the speaker stated [8.2] : "Forget Ethanol – – – – . Another alcohol, Butanol is a much better renewable fuel." A great deal of research has been invested on this fuel, and it holds much promise, with many key advantages: less corrosive; easier to blend; and can be moved via pipeline. Its major disadvantage is cost. A fermentation process is used, but is still under development. Dupont has ben working on the "butanol bug" since 2004. Although the final feedstock(s) remain to be defined, the initial pilot plant and commercial plant will be based on wheat, as there is a surplus of wheat in the U.K These plants[8.3], will be located at Hull, England, and will be 5K and 110M gallons per year respectively. This is almost an oxymoron statement, as one does not commit to the size of a commercial plant until the pilot plant has operated for some time and the final process design is pinned down. Could there be a bit of hype in this statement? A suggestion that there are serious problems with this development was noted[8.4] in a blog that asked the question "Why these industrial giants are bothering with bio butanol when it is at best an inadequate improvement on the scam of corn ethanol - - -." This writer also cited a report[8.5], by Robert Rapier, that concluded that "biobutanol is not remotely at the level the hype implies." Both writers[8.4, 8.5] agree "though superior to corn ethanol, biobutanol's inefficient, energy-wasting fermentation process makes it at least 10 years from being a useful fuel source."

Hydrogen.

For a summary of hydrogen production options[9.0], see the cited reference. The writers noted that hydrogen can be released from water and hydrocarbons, but it always requires energy to do so - more than can ever be recovered by using the released hydrogen as fuel. "So for all practical thermodynamic purposes, hydrogen is an energy storage medium, not a source of energy."

Promoters of H_2 frequently claim its only emission is water vapor, from the fuel cells employed. This is duplicity of the highest form. For example, in a 2004 paper[9.1] the writer, noted that "given current technology switching from gasoline to H_2 powered fuel cells would greatly increase energy consumption – – – nearly double greenhouse gas emissions."

One might cite several routes to get to H_2 supply.

(1) <u>Direct Electrolysis</u>. H_2 could be extracted from water via electrolysis. This would take a major amount of electricity and is a most expensive route to hydrogen. Further, any emissions caused by this new demand for power would need to be allocated to hydrogen. See NOTE 12 for the reaction.

About 4% of global hydrogen production is by this process.

(2) <u>Indirect Electrolysis by thermo-chemical cycles</u>. In theory H_2 could be extracted from water via electrolysis by thermo-chemical cycles. One such route is the Hybrid Sulfur Process[9.2]. Again see NOTE 12 for reaction details.

(3) <u>Steam Reforming</u>. Almost all H_2 produced in the world today is via this process. The raw material for this process is inevitably natural gas. It has been estimated[9.1] that about 15 trillion cubic feet of gas would be needed, annually, to produce the necessary H_2 for the United States to power the vehicle fleet. This would boost the consumption of natural gas in the U.S. by about 66 percent. Even today gas supply continues to call for major imports from Canada, and even from the world. Further, any

emissions caused by this new demand for H_2 would need to be allocated to hydrogen.

(4) <u>Solar-themal processes</u>. It is possible a relatively new process, but unproven could emerge, or it is possible a brand new and essentially unknown process could emerge. See NOTE 12 for ideas.

<u>Summary</u>. One can argue there are either cost or technology problems or both on hydrogen production. As hinted above, there are also major problems with H_2 distribution, storage and use. The current fuel cells were developed for the space program and may not be optimized for autos. There is also the issue of cost. So Hydrogen is still viewed, by this writer, as a long shot as a replacement for gasoline.

Solar Photovoltaic.

I will cite two writers on this subject.

(1) <u>The first writer</u>[10.1], Robert Bryce, gives his own experience on his home in Austin, Texas. His unit cost him $7445, after a subsidy of $15,000, and saved about $386 in 2007. It supplied 31 percent of his power. The payback, based on 2007 results, would be 19 years. If it had not been subsidized by the city, this value would be 58 years. In short this writer sees solar as having a very long way to go. He calls solar "The 1Percent Solution."

Bryce also cites the U.S. Government 2006 projections for 2030: solar generation - 5 bkwh; and for total generation - 3351 bkwh. This is only 0.15 percent.

(2) <u>The second writer</u>[10.2], asks why is solar so expensive? Total cost today averages $8.25 per watt, of which the solar panels would amount to $3.75 per watt. For the 3,240 watt unit in Austin, installed cost would have been $26,730. Hence the solar panel plus installation are both very expensive.

He also addresses the potential for major improvement, analogous to that seen in the chip industry. He noted that the rate of improvement is nowhere near the chip history. And it will not be. While computer chips are getting exponentially smaller and hence will need less raw material, as the same opportunity for size reduction does not exist with solar panels. Prices for solar panels haven't declined over the past five years because most manufacturers are paying multiple times more for raw materials than in 2002. Further with the current fuel and power inflation it seems rather unlikely that cell costs will come down.

Conclusion on AE.

The above is a sampling of key, prominent areas of AE. In spite of its costs, its embryonic status and the need for major research breakthroughs, the need for subsidies, its low availability (and hence need for backup generation capacity), in some cases the low liquid fuel contribution, the remote locations (and hence need for transmission capacity), the various forms of AE can make a contribution. But can they be the solution to our current crisis? The previous inputs would suggest not. Others agree. For example Robert Bryce has concluded "it should be obvious that the U.S. cannot give up its reliance on oil. "Nor can it give up its reliance on coal and natural gas. "Put simply America will be using fossil fuels for decades to come."

6.4 Emerging Strategies II: Focus on electrified transportation.

Introduction.

At least two, equally crucial elements are needed for this pathway to become relevant. Clearly a superior battery—other than the lead acid unit used in autos and golf carts—will be needed. In addition to coming up with such a battery huge amounts of new electrical generation capacity would be needed.

• High Energy Density Battery R & D.

As a former employee at a national lab in Canada, and as a project business manager of key research projects at a major chemical company in the U.S., I do not want to minimize the role of R&D. The job of a research scientist has never been easy. It requires a rather incredible knowledge of a specific field, but equally incredible tenacity to progress a never ending array of details. Then as the current pathway fails, to pick oneself up and start over again. Never has this picture been more true than on specific battery research.

Here I will limit my comments to the Sodium Sulfur and the Lithium ion batteries.

(1) <u>Sodium Sulfur battery</u>. One area that looks very promising is the emergence[11.1-11.2] of the Sodium Sulfur battery (NaS, where Na is the chemical symbol for Sodium and S for Sulfur). Note that this approach

was pioneered by Ford Motor Company, over 40 years ago, for the auto application. Their approach used a beta-alumina ceramic membrane as the critical separator. It was also researched extensively by my old chemical company[11.3]. One key problem for this battery in autos was a very high operating temperature of 350 ºC, hence a key safety problem.

However, the NaS unit was brought to the demonstration stage for electric utility power storage, by a Japanese company and American Electric Power (AEP). Many NaS batteries are in use in Japan, and AEP has tested a 1,200 kw unit, with plans to add a unit twice that size. Another utility is planning on a 5,000 kw unit.

A slightly different application[11.4] involved an installation at a major bus company. This is the first installation on the *customer side* of the power meter in the United States. This installation that used electric mto drive three compressors, used to refuel natural gas busses. This battery is capable for providing one megawatt for up to seven hours a day. It permits buying power at off peak times, plus cutting back a shift in operations.

(2) Lithium ion battery (Li-ion). First note the distinction between the Lithium and the Lithium ion batteries. Most Lithium batteries can not be recharged, so for this essay our interest is focused on the Li-ion unit. While much of the development has been for small applications such as laptops, interest is now moving rapidly to units large enough for hybrid vehicles or full electric vehicles. Surely, developments in the smaller units may well have a spin off on the bigger units.

Recent developments[11.5] on the Li-ion battery—where the initial work is 100 years old— are surely encouraging. There are many major organizations active in this field, and many technical developments emerging. See NOTE 13 for details.

Conclusions. These two areas surely look promising. While it is way too early to decide specific applications, a few observations may be worthwhile.

(i) NaS Unit. This unit looks like s fit for standby power. It is rugged looking and reliable looking. It would seem that it could be built just about as large as desired. However, it does not appear to be a fit for the vehicle market, primarily due to its high operating temperature.

(ii) Li-ion Unit. This looks good for vehicles and clearly is getting all of the attention today. With the new technologies reported above, the previous problems with fires and explosions should be a thing of the past. Indeed optimism for this unit is now very high. At least five manufacturing plants are planned: four in Michigan and one in Kentucky. One forecast has suggested [11.13] a growth from $700M in 2008 to $3.2B by 2012

Surely more hard data is needed on both of these units, particularly on cost, performance and lifetime. However, it would appear many incentive dollars will be used in this area. It is important to remember that inventions cannot be dictated by government fiat, and historical development times for these and other above batteries stand as testimony to the challenges in this field.

• Power Support.

(1) <u>Fossil Fuel based</u>. If an acceptable battery emerges, it just might have to be supported by other types of power plants than nuclear. For any fossil fuel plants built to support auto/truck fleet electrification, we would clearly need the fuels to fire such plants, not an easy pathway for environmentalists/ warmers to take.

(2) <u>Nuclear</u>. If an acceptable battery emerges, it just might see the rebirth of nuclear power. I believe this is the spirit that Senator McCain called for 45 new nuclear plants, during the past election. However, as one who spent part of his career on trying to get a Michigan nuclear project rolling—about half a dozen tiny assignments, plus dozens of letters to the local newspaper only to see it ultimately converted to a natural gas plant— I am convinced a program to build 45 nuclear plants will be extremely difficult to get underway, and even more difficult, if not impossible, to implement.

6.5 Emerging Strategies III: Focus on Fossil Fuels.

• Oil and Natural Gas.

We can do much more to improve domestic oil and natural gas supply. *No way, the warmers scream.* Indeed, for any help from oil and gas, we will need to get away from the anti oil crowd. This crowd—whether it is outer continental shelf (OCS) oil or the ANWR—inevitably bad-mouth such initiatives as providing only a tiny amount of energy and will never help our situation. In reality it is their ideas on AE that will net only tiny amounts of energy. Their bad mouthing on ANWR and the OCS includes their claim that such oil development would only cut [2.6] three cents off the price for a gallon at the pump. However, if one assumed ANWR would yield 10 B barrels of oil, over a 25 year life, we are talking about 1 MBPD. There would be a cut of the same size off of our global oil call. And only three cents? Please!

While U.S. oil production peaked over thirty years ago, it is now seeing a bit of a comeback, due primarily to the huge oil find in the Bakken Shale of MT, ND and SK. And natural gas[2.7] is enjoying a bit of a boom. In 2009, gas production surged[2.8] due largely to unconventional gas (UCG) resources such as the Barnett Shale, that hadn't been tapped in the past. This UCG resource—shale gas, tight sands gas and coal bed methane—has shown a rather surprising increase[2.9], especially in view of recent dire predictions regarding North America's increased dependence on imported LNG. Note that the EIA projects unconventional gas will represent about half of total U.S. production by 2012.

While this increase in UCG speaks volumes about drilling technology and geological savvy in managing these resources, it does not speak well of the state of conventional gas resources. This includes the fact[2.10] decline rates for gas wells have almost doubled over the past ten years. Perhaps more ominous is the well known dramatic decline rates for shale gas. All of these inputs suggest we will soon be looking strongly at LNG again

• Coal.

This fuel is undergoing massive expansion in China, India and elsewhere. And we are seeing some of that here. But we can do much more with coal. Coal use can be expanded today not only for power, but for gaseous and liquid fuels too. For coal, the resource base is almost unlimited. The U.S. has been called the "Saudi Arabia of coal." Perhaps we need a program[2.4] here on coal, analogous to the Canadian effort on the Athabaska Tar Sands, although that would be a highly controversial, perhaps even an insane idea.

Note that many additional comments on this fuel follow in Section 5.

• Shale Oil.

And we also have vast reserves of shale oil. This is more difficult and costly to process than Alberta tar sands or coal, but none the less a significant potential.

• Overall Contribution and Impact.

Now activists argue it would be years before any supplies emerged from this fossil fuel strategy. This is partially true in the sense that new off shore platforms, coal based plants, pipelines, and oil shale units have a long construction span. But price relief can come much quicker than that.

All that is needed is for the speculators of this world to become convinced that we are finally going to put our energy house in order. To achieve that we would have to launch a dedicated, definitive and aggressive program to maximize use of all domestic energy resources, with significant effort on fossil fuels. This would include ANWR and OCS oil. It would include new coal fired power plants, and some progress on coal gasification units and coal liquefaction plants. Once the speculators become convinced we are going to **reduce our call on global oil**, starting soon, but extending over a set of future dates certain, the price of oil would start to come under control.

Today, we don't have any supply security on liquid fuels, and we don't have any price security on such fuels and to a lesser extent on power. Our fossil fuels are the way to such security.

6. 6 Additional Comments on Coal and Related Subjects.

One cannot expect the warmers/environmentalists to salute the fossil fuel focus plan. As with the nuclear option, opposition will be loud, massive and entrenched. Indeed, this could well be the fight of the century. However, we don't have a century. In fact, we may possibly have less than a decade.

The Climate Change Situation.

(1) <u>Background</u>. I have followed this issue extensively for 20 years. I have studied it rigorously and attended many conferences. I have subscribed, at one time, to dozens of scientific journals, and used the university libraries in Houston to complement this input. I have used the Internet voraciously. And I have written on this subject. My recent comments[12.1 -12.4] have focused on what I call the key witnesses for the defense of the skeptical perspective. These witnesses include:

(i) <u>Key Non-Scientists</u>, but authors with some unique perspective on this issue;

(ii) <u>Distinguished Veterans</u> (*DVs*), mostly scientists, mostly retired, but with incredible accomplishments. I use italics here on the *DVs* to differentiate these veterans from those of the military. These are veterans from the research community, some with emeritus in their title. I have developed a listing of over 60 such scientists. In a recent paper[12.3] the views of eleven *DVs* were noted: four experts on hurricanes, three on physics and four on various aspects of food production. In another paper[12.4], the views of 17 physicists, all skeptical on GW, were noted.

(iii) <u>Others</u>, including active scientists, TV Meteorologists and State Climatologists.

(2) <u>Conclusions on the Climate Change Situation</u>. In a recent paper[12.3] the views of the above witnesses have been noted. Their lifetime publications and comments give the nature of their views on the GW issue: all skeptical. There are simply too many highly educated, high-horsepower individuals[12.5] -- who are concerned that we have not diagnosed the climate scene completely or correctly -- to ignore their views. As an example of this inadequate or incomplete diagnosis, consider the most recent inputs[12.6] from Bjørn Lomborg. He argues that, while GW may bring an increase in heat-related fatalities, it would also lead to a far bigger reduction in cold-related deaths. There are six times the number of cold-related deaths in Europe than heat-related deaths, while for the globe the corresponding numbers are 1.8 million and 400,000. So GW would result in a net saving of 1.4 million lives by 2050 on this vast difference alone. This writer thus concludes that "if our starting point is to prove that Armageddon is on its way, we will not consider all of the evidence and will not identify the smartest policy choices."

Today, the proponents of this issue are a heterogeneous mix. This includes the behind-the-scene organizers; it includes many who are intimidated by the fear of job or funding loss; it also includes many fellow travelers who are riding the political winds, or feel the need to be politically correct; and it includes many who have been simply brainwashed on the issue. A slight variant of this last category was noted[12.7] recently, as those who have accepted GW simply because "endorsing global warming just makes their lives easier."

Consider the behind-the-scene organizers[12.8] and check out the Grassroots Partners and Advisory Committee members for "We." This organization -- *We Can Solve It* (initiated in part by Albert Gore Jr.) -- has come out with some rather professional, high-power TV ads aimed at convincing the viewers that there is a *climate crisis* and that it is urgent and solvable. However, the very *best* one can say about the GW issue, and the need to move to a carbon-constrained world, is that it is *premature*. The very *worst* one can say is that it's a *fraud*.

The Environmental Situation in General.

I have followed the environmental issues in general for over 40 years, and for the coal area, in particular, for over 30 years. My position in this area is that we can live with this fuel. One might ask: what are my credentials for taking such a position? This position is based on inputs from three areas: (1) emissions control improvements, (2) over 75 years of individual exposure to coal, oil and key chemicals; and (3) experience obtained via service on The National Coal Policy Project over the period 1977 to 1979.

(1) **Emissions Control Improvements**. As noted above, the key pollutants from coal-fired power plants have been controlled. On balance, we've achieved much cleaner generation of electricity from coal since the Clean Air Act of 1970, as two additional sources also note.

Peter Huber[13.1] notes that fossil fuels extract more power from less of the Earth's surface than AEs "and are therefore greener," adding that *soft energy* sources "are horribly land intrusive." Huber agrees that much progress has been made with some of the worst forms of pollution, such as big smokestacks. He warns, however, of the danger of doing less and less with more and more regulations. Hence, he advocates "visible green objectives," such as wilderness conservation and park creation. As noted on its cover, his book may make you examine your assumptions, which is something that is sorely needed today.

Bjørn Lomborg[13.2] paints a picture of a world where human welfare is improving in just about every way one might measure it. He notes that the "achievement of dramatically decreasing concentrations of the major air pollutants in the western world ... is amazing by itself." And this has been achieved while the economy has increased dramatically. Coal-based power plants have experienced tremendous improvement in emissions control. With such improvement in the environment in general, and with coal units in particular, it is strange that we hear so much bad news about the environment. Could it be that environmentalists often lie? Within this context, Lomborg demonstrates many ways in which professional environmentalists do indeed play fast and loose with the truth.

(2) **Individual Exposure to Coal, Heavy Oil and Critical Chemicals**. In-spite of much progress on the pollution front, the public is still concerned about a heavy reliance on coal and heavy oils. As a means to soften such concerns, I will offer a view of my trip through life, particularly as it has been involved with coal, heavy oil, and other potential environmental

problems. Now I know this represents only a sample of one; and I don't want to present this odyssey as any sort of scientific proof that these commodities have never had major problems in the past. Rather, I present it as *food for thought* and to raise the possibility that certain concerns about coal may have been overstated.

(i) <u>Saskatchewan</u>. I start this trip early in my life in the city of Saskatoon.

(a) <u>Home heating</u>. I was born into a coal-fired home, and spent my first dozen years living with coal. Everyone had a coal bin, and you had periodic coal deliveries. And there was always a certain amount of coal dust. Over this period, we converted initially to a coal stoker—which was a major improvement in convenience—and later we went to fuel oil and eventually to natural gas.

(b) <u>Coal fired trains.</u> Our home was only one short block from the Canadian National Railroad tracks heading to the downtown station, and other destinations. The coal fired engines, over a dozen or more a day, either slowed down as they crossed the river and entered downtown, or speeded up, as they left. There were undoubtedly emissions from these engines, plus soot and ashes. I spent my first six to seven years with this situation.

(c) <u>Source of electricity</u>. I was also born into a city with a coal fired electric system and spent my first 22 years living with this system. The local utility had two, large coal fired steam-electric stations. One of these was just across the river, about ¼ of a mile away. The newer unit was about five miles southwest. And there were undoubtedly emissions from these plants.

(d) <u>Local refinery</u>. This refinery was about as small as a refinery could get, but one always knew it was around. It was also located about five miles to the west; yet when the west wind blew (which was most of the time), the aromas would drift our way. And one could also see the flare at work.

(e) <u>What does one do with a used coal bin</u>? When I was 18 I took a summer job in construction. The first task was on a project where a building was being renovated, and a coal bin had to be torn down. Naturally, such a job was given to the *rookie*. This basement room had plaster walls and ceilings, complete with supporting lathing. And coal dust was everywhere. It took about a week to complete this job, and each day I came home as black as one could get. Of course *OSHA* type rules did not exist back then. I was not let in our house until I stripped down to my briefs, and was as hosed down as one could get. I could clean the outside of my body, but doubtful on the inside.

(f) Local asphalt plant. For two summers, I had a job as an inspector at the city's asphalt plant. There were at least two problems at this plant: dust from the hot gravel and tar-oil vapors.

(ii) Ontario. After graduating I went to work in Ontario, with a major oil company.

(a) Refinery Design. This job included work for refineries in Calgary, Vancouver and Norman Wells, Northwest Territories. It was clean work, but our offices were across the street from the largest refinery in Canada.

(b) Petrochemical Plant Startups. The first startup was on a detergent alkylate plant. Everything that could go wrong with a new plant did; and I was involved in this startup, for almost a year. It used benzene as one of its raw materials, a known toxic substance. Before World War II, this chemical was made from coal tar and was recognized as a bad actor. Several precautions were taken, but there were bound to have been trace benzene leaks all during this time.

The second startup was on a heavy oil steam cracker for ethylene, propylene and butadiene production. Fortunately this startup went fairly smoothly.

(iii) Michigan. Next, I worked for a huge chemical company in Michigan.

(a) Steam and power system. Coal fueled this plant, and the first environmental problem of concern was not chemical but coal dust. We had just moved into a new apartment in 1960, when we discovered our car and our outdoor window sills would periodically turn black. If we ran a damp sponge over the inside window sill, it would also turn black. This was enough to leave this company, if relief could not be obtained.

This plant needed electricity for many pumps, compressors, mixers and for the electrochemical manufacturing of chlorine. It also needed steam for many heat transfer functions. Both electricity and steam could be produced via a co-generation process in which this company was the world's leader. But the power plants were old, with no stack-gas treatment. When I complained about the problem, I was told be patient, that relief was coming. We lived with this situation for the better part of a year before it arrived.

(b) Chemical exposure. I spent brief amounts of time in an olefins – refinery complex, a glycol ether brake-fluid plant, a polystyrene plastics plant, a styrene butadiene latex system plant, and finally a herbicide production plant. So I was close to some serious safety and health problems.

In 1962 the military came to my company, and ten others, to make Agent Orange, a mixture of two commercial herbicides[14]: 24D and 245T. My group had a project in this area on scheduling and inventory control. One of the herbicides, if it was not manufactured properly, could contain an impurity, a chemical called dioxin. Our scientists developed new analytical technology that could measure dioxin to below the parts per trillion level. Because of this new capability, they found that dioxins were all over our society, as a result of burning many substances. They were found in auto mufflers, cigarette smoke, wood soot and other places, including chemical plants and paper mills.

(iv) <u>Texas</u>. We had just moved into a new home in February 1982, when within a month we discovered our car and the surface of our pool would turn yellow. This was deja-vu all over again, as in the case of our Michigan experience, except the residue was yellow, not black. My immediate reaction was that a sulfur plant, perhaps some 20 miles away, must have blown up. But I was wrong. Turns out it was pine tree pollen.

Over the balance of my career, I was deeply involved with hydrocarbon supply, coal and lignite planning, power supply planning, and coal gasification work. This led to visits to coal/lignite mines in North Dakota, Wyoming and Texas, plus many power plants throughout the U.S.

(v) <u>Summary on personal exposure to coal dust, tar and critical chemicals</u>. I believe one can conclude that I have had more than my share of exposure to coal dust, tar and critical chemicals. I am now 77, and going strong. Again, I realize that this is only a sample of one; but it is food for thought. Can the portrayal of hideous problems associated with coal be just a bit over-stated?

(3) ***The National Coal Policy Project - 1977 to 1979***. There were also environmental concerns, back in the 1970s, on the further use of coal and lignite. As a result, our corporate energy manager, along with a leading environmentalist, initiated what was called the National Coal Policy Project, under the auspices of Georgetown University. Here, several environmentalists met with several industrialists to try to define a future pathway for coal that would be acceptable to both camps. This was an excellent idea, but it was an incredibly difficult and frustrating project. Indeed, it was decided early on "to leave the task of making projections to others." However, some 200 recommendations were made[2.5]. I don't want to dwell on these, however, as they are 30 years old. Instead, I want

to recount my impressions of this activity. In doing so, I note that I was on the coal transportation sub-committee, and that these impressions come primarily from that activity.

My first impression was that the environmentalists were better prepared for these meetings, as if they had spent all of their time getting ready for them. In contrast, all of the members of the industrial side would go back to their regular duties and have very little time to do much homework.

My second impression was that we could never get the environmentalists to make a list of the problems and define their priorities, at least as far as I could see. Later, I came to the conclusion that they would *never* provide such a list.

My final impression was that the attitude of some members of the environmental movement "scared the devil out of me." They were so intense, so certain of their cause, and so socialistically oriented, such that *in their mind* no discussions were really needed. They seemed to be saying "get us elected and get out of the way."

Today I hear what all the activists and politicians have to say about coal; and I'd swear they are the same people I met 30 years ago. They are so intense, so certain, and so socialistically oriented, that *in their minds*, no discussions are really needed. They just seem to be saying "get us elected and get out of the way."

6.7 Conclusions.

Today, our country is in the most serious energy situation it has ever faced. It could make the gas lines of the 1970s look like kid stuff. Here, I have argued, directly and indirectly, the following points.

• While the various forms of AE can make a contribution, they will not be the solution to our current crisis. Robert Bryce, in Section 3.5, has stated it well: "It should be obvious that the U.S. cannot give up its reliance on oil ... nor can it give up its reliance on coal and natural gas." Put simply, "America will be using fossil fuels for decades to come."

• The pathway to price and supply security is to **reduce our call on global oil** via a substantial boost in our conventional supplies of domestic oil and gas, a start on coal liquefaction, a start on shale oil, a dramatic improvement in electric vehicles, including hybrids, and the power system to support such a move (a power system based on coal-fired plants, coal gasification units, and to a lesser extent, nuclear units), and a boost in AE and on energy efficiency and conservation.

This is not a call for another "Project Independence", but one of showing the world we can *and will* manage our energy situation.

• AEs are not the pathway to the desired price and supply security. They are useful in specific areas; but in no way are they the overall solution. Perhaps the most exciting activity in this area is the development of high energy density batteries. The least exciting activity, indeed the almost *criminal* activity, is the incredible support for corn-based ethanol.

• The climate change situation, in regard to fossil fuel use, is manageable; and the need to move to a carbon-constrained society is premature at best.

• Even the need to move to a "clean coal" society must be carefully reviewed and more realistically defined.

• The environmental situation, with respect to coal utilization, is manageable, which judgment based in part on my own lengthy exposure to coal, and in part on the tremendous improvement in emissions control over the past 30 to 40 years.

It will take leaders of great courage and wisdom to develop and manage such a program. It will take leaders of intelligence and ingenuity to get the *Phoenix* -- our *country* -- off the ground and flying in the right direction. Unfortunately, at a time when we need such leaders, they appear to be in short supply. Let us hope they emerge very soon.

References and Notes.

(1) The movie The Flight of the Phoenix (1965) included actors James Stewart, Richard Attenborough and Hardy Kruger. It was a tense and character-driven film about a small but diverse group of men struggling to overcome the adversity of a harsh and deadly environment, who also had to come to grips with each other and the character strengths and flaws inherent within themselves, in order to complete a nearly impossible task, one that would determine their very survival.
(2) Oil and Natural Gas References.
(2.1) Simmons, Matthew R., *The Oil and Gas System is Sick*, presentation at The Commercial Club of Boston, February 11, 2009. Simmons passed away suddenly

on August 8, 2010. He was widely regarded for his analysis of the oil industry and his investment council.

(2.2) Mouawad, Jan and Henriques, Diana R., *Why is Oil So High? Pick a View*, The New York Times, June 21, 2008.

(2.3) Simmons, Matthew R., *How Mature are the World's Super Giant and Giant Oil & Gas Fields and are they Still Important*, presentation, Pioneer Oil Producers Society, Houston, March 16, 2009.

(2.4) Will, George F., *Opposition to ANWR drilling? It's collectivism in drag*, Houston Chronicle, December 16, 2005.

(2.5) Evans Thomas W., *A strong case to be made for suing OPEC profiteers*, Houston Chronicle, June 20, 2008. Evans is a former advisor to Presidents Ronald Reagan and George H. W. Bush.

(2.6) Based on a Real Clear Politics blog: *McCains Half Hearted Oil Solution*, June 20, 2008.

(2.7) Kohl, Kieth, *Natural Gas Production*, Energy and Capital, September 15, 2008.

(2.8) Fowler, Tom, *A role for coal, new drilling*, Houston Chronicle, February 10, 2009.

(2.9) Steinhubl, Andrew, et al, *Unconventional resources to keep pivotal supply role*, Oil & Gas Journal, January 26, 2009.

(2.10) Letourneau, J., *Where is the North American Natural Gas Market Headed*, Seeking Alpha, June 15, 2007.

(3) Coal references.

(3.1) Johnson, Keith and Steven Chu: *'Coal is My Worst Nightmare'*, The Wall Street Journal, as reported in http://blogs.wsj.environmentalcapital - - -, December 11, 2008.

(3.2) Deutch, John and Moniz, Ernest, *A Future for Fossil Fuel*, Wall Street J., March 15, 2007.

(3.3) *The Future of Coal*, MIT News Release, March 14, 2007. This news release is for the report: *Future of Coal - Option for a Carbon Constrained World*.

(3.4) National Coal Council, *Coal: Americas Energy Future*, as reported in Industrial Environment, May 1, 2006, and as posted in Goliath Business News.

(3.5) Center for Strategic and International Studies, Georgetown University, *Where We Agree - Report of the National Coal Policy Project*, Westview Press, Boulder, CO, 1978.

(4) Political positions on climate change.

(4.1) Graham-Felsen, Sam, *The World Beyond Iraq*, Organizing for America, March 19, 2008. See: http://my.barackobama.com/page/community/post/samgrahamfelsen/gGBFrl.

(4.2) Herszenhorn, David M., *After Verbal Fire, Senate Effectively Closes Down Climate Change Bill*, New York Times, June 7, 2008.

(4.3) Daly, Brenda, Elshami, Nadear, *Pelosi, Reid Call on Bush to Support mandatory Limits on Greenhouse Gases*, Speaker Nancy Pelosi press release, September 28, 2007.

(5) Wind energy.

(5.1) Steffy, Loren, *Wind Whispers of Enron*, Houston Chronicle, June 2, 2008.

(5.2) Bradley, Robert L., *Corporate Social Responsibility and Energy: Lessons from Enron*, Institute for Study of Economics and the Environment, Lindenwood University, April 2008.

(5.3) Horner, Christopher C., *The Politically Incorrect Guide™ to Global Warming and Environmentalism*, Regnery Publishing Company, Washington, DC, 2007.

(5.4) McQueen, Ian, Sing a Song of Turbines, Telegraph-Journal, Canada East, June 20, 2008.

(5.5) *Formal study needed into health effects of wind turbines, doctor says*, The Canadian Press, April 23,2009.

(5.6) Tyner, Sr., Gene, *Net Energy from Wind Power*, Minnesotans For Sustainability, Jan. 2002.

(6) Biofuels in general.

(6.1) Anthrop, Donald F., *Biomass potential*, Oil & Gas Journal, September 5, 2005.

(6.2 Radler, Marilyn, *Ethanol and oil markets*, Oil & Gas Journal, April 3, 2006.

(6.3) Gardner, Timothy J., *"New ethanol" to face crunch under a Chu DOE*, eleconomista.es, November 12, 2008.

(6.4) Lammers, Dick, *Can ethanol be the fuel of the future?*, Houston Chronicle, February 15, 2009.

(6.5) Philpott, Tom, *New Energy Secretary Chu: Big proponent of cellulosic ethanol*, BioDieselNow, December 17, 2008.

(7) Ethanol from corn.

(7.1) Carrol, Joe and Parker, Mario, *Ethanol Bust Makes Losers of Bush, Gates, D. E. Shaw* (Update2), Bloomberg.com, November 21, 2007.

(7.2) Delaney, David, *Why Ethanol Can't "Solve" the Fuels Problem*, Peak Oil and the Fate of Humanity blog, July 2, 2005.

(7.3) Rapier, Robert, *Mythbusters: Ethanol and Foreign Oil Displacement*, The Oil Drum, as reported on The Intelligence Daily, August 8, 2008.

(7.4) Wald, M. L., *Alternate Fuels: All Gallons Are Not Equal*, New York Times, May 28, 2006.

(7.5) Editorial, *Risky lans and ethanol*, Oil & Gas Journal, January 26, 2009.

(7.6) Streifeld, David, *Uprising Against the Ethanol Mandate*, New York Times, July 23, 2008.

(7.7) Boudreaux, Don, *Issue: Ethanol*, America 2012: Business and Media Institute Special Report, Undated.

(8) Biodiesel and Biobutanol.

(8.1) Johnston, M. and Holloway, T., *A global comparison of national biodiesel production potentials*, Environmental Science & Technology, 41 (23), 2007.

(8.2) Verma, R. P., Buanol - A possible Alternative Energy Source, International Symposium on Biofuels, September 25, 2007. See http://petrofed.winwinhosting.net/uploa/4_Verma.pdf.

(8.3) Campoy, Ana, *Betting on a Biofuel*, Wall Street Journal, June 30, 2008.

(8.4) Submitted by New Energy News Blog, *IS BIOBUTANOL THE BIOFUEL? NOT YET*, Energy Stocks Blog, July 2, 2008.

(8.5) Rapier, Robert, *The Problem With Biobutanol*, R - Squared Energy Blog, Jun 12, 2007.

(9) Hydrogen references.

(9.0) The Hydrogen and Fuel Cell Investor, © 1999 - 2010. See: http://www.h2fc.com/technology/hydrogen/production.shtml.

(9.1) Anthrop, Donald, *Hydrogen's Empty Environmental Promise*, Cato Institute Paper No. 30, December 7, 2004. Anthrop is a professor emeritus of environmental studies at San Jose State.

(9.2) - (9.5) See NOTE 12.

(10) Solar photovoltaic references

(10.1) Bryce, Robert, *Gusher of Lies - The Dangerous Delusions of "Energy Independence"*, PublicAffairs™, 2008.

(10.2) Chernova, Yuliya, *Shedding light on Solar*, Wall Street Journal, June 30, 2008.

(11) High-energy density batteries.

(11.1) Priore, Suzanne, *AEP Dedicates First Use of Stationary Sodium Sulfur Battery*, American Electric Power, September 23, 2002.

(11.2) Davidson, Paul, *New battery packs powerful punch*, USA Today, July 4, 2007.

(11.3) From 1966 to 1978, I was project business manager for Membrane Systems at The Dow Chemical Company. One of several projects was a research effort on a NaS battery. At the same time, Ford was developing this chemical cell based on a ceramic membrane. The Dow approach was based on a specialty glass in a hollow fiber configuration. Both Dow and Ford invested many years of research effort and dollars, along with significant government support. Neither of these efforts reached the light of commercial reality. However, the Ford effort probably contributed substantially to the Japanese results.

I periodically compared notes with a colleague in Dow who had similar responsibilities on a Magnesium dry cell, then under development. I believe he had been on that task for about 20 years. A major part of this time was devoted to quality control. Note that for the future batteries of this world, the quality control needs would surely be exponentially greater than for this dry cell example.

(11.4) *MTA LI Bus and NYPA Install First Sodium Sulfur Battery Energy Storage System in State*, January 9, 2009, as reported in http://www.nypa.gov/press/2009/090109a.htm.

(11.5) Buchmann, Isidor, *Is lithium-ion the ideal battery*, Created: April 2003, last edited: November 2006. See http://batteryuniversity.com/partone-5.htm.

(11.6) to (11.12) See NOTE 12.

(12) Global warming.

(12.1) Westbrook, Gerald T., *The Skeptics on the Global Warming Issue: The Distinguished Veterans*, IAEE Newsletter, 4th quarter, 2005.

(12.2) Westbrook, Gerald T., *Global Warming: Who to Believe?* AICHE, April, 2007.

(12.3) Westbrook, Gerald T., *Global Warming: Witnesses for the Defense of the Skeptical Perspective*, Energy Tribune, May 29, 2007.

See www.energytribune.com/articles.cfm?aid=500.

(12.4) Westbrook, Gerald T., *Global Warming: Witnesses for the Defense of the Skeptical Perspective - Physicists*, IAEE Energy Forum, 3rd quarter, 2008.

(12.5) Westbrook, Gerald T., *Warming debate needed*, letter, Oil & gas Journal, November 3, 2008.

(12.6) Lomborg, Bjørn, *Global warming will save millions of lives*, The Telegraph, March 12, 2009.

(12.7) Lindzen, Richard, *Climate Alarm: What We Are Up Against, and What to Do*, Heartland Institute Conference, March 8, 2009.

(12.8) See: http://www.wecansolveit.org/. In particular select: *About us*, for access to Grassroot Partners and Advisory Committee members.

(13) <u>Coal plant emissions</u>.

(13.1) Huber, Peter, *Hard Green - Saving the Environment from the Environmentalists*, Basic Books, New York, NY, 1999.

(13.2) Lomborg, Bjørn, *the skeptical environmentalist - Measuring the Real State of the World*, Cambridge University Press, 2001.

(14) <u>The commercial herbicides were</u>:

· 24D or more precisely: 2,4-D. The chemical formula is 2,4 - dichlorophenoxyacetic acid and

· 245T or more precisely: 2,4,5-T. The chemical formula is 2,4,5 - trichlorophenoxyacetic acid.

7. A Critique of the Environmental Movement

7.1 Introduction.

As I noted in the Preface, there is much propaganda in the world today, and much of it focuses on the environment. The bulk of the propaganda on the environment comes from the Left. The Democrats will, of course, argue the exact opposite. However, the Left have out spent the Right ten to one on 527 advertisements. And the left is prone to such efforts as the cover[0] on a recent Time magazine depicting a tree planting analogous to the flag raising on Iwo Jima. The text on this cover was: "How to Win The War On Global Warming", and included a green border rather than their traditional red. This effort raised the wrath of many WWII veterans.

• The Left Will Use Absolutely Anything.
This was the title of Chapter 13 in my first book. Seldom has the Left been more egregious in the use of hype and distortion than with the following charge against the City of Houston, used during the 2000 presidential election. The Left became rather desperate and were looking for anything to throw at the then Texas governor, much as they were desperate during the 2004 election. Back in 2000 a Democrat spokesperson, Jennifer Lazlo-Mizrahi, amongst others, "launched an assault on the City of Houston saying 'we need to start letting people know, for example, that Houston is a *filthy, smoggy, disgusting city*'[1]."

The basis for these charges was that Houston had replaced Los Angeles as the most polluted city in the country. If true the implication was that George W. Bush was responsible for this situation and guilty of being an anti-environmentalist. Some comments follow.

Here the Democrats were striving to paint George W. Bush as guilty of terrible environmental leadership and stewardship. Not only did they ignore the accomplishments of Texas over the past 10 to 20 years, but also ignored the fact that Bush, as governor of Texas, had zero responsibility and zero control over the Houston air quality in general, and any smog there in particular.

If there was any one politician that had some responsibility and influence on this matter, it would have been the mayor of Houston, Lee P. Brown. A Democrat, his influence was that of the bully pulpit—as the EPA, and the state analog of the EPA, were in charge and monitored emission releases and air quality very carefully, and continued to push Houston forward on the one criteria pollutant, ozone, that was out of compliance—but nowhere was Browns name seen connected to this problem.

Where the Mayor's bully pulpit was needed was to defend the city from the unfair and scurrilous attacks coming from Washington and New York. His cohorts—in other democratic offices in Houston and in Washington—also did nothing to correct this situation. In contests where image, and the perceived quality of life of each city, was going to be the dominant factor, these politicians did zero to refute these spurious charges.

• This is Park Country.

It took Elyse Lanier, the wife of the former mayor, Bob Lanier, and chairwoman of the Houston Image Group, to set the record straight[2]. "Houston is an amazing city, with great museums, theaters, renowned colleges, family attractions..." She went on: "I know from personal experience that once we get people here to visit ...they are pleasantly surprised. "They didn't know Houston was so green, had so many restaurants, theaters, museums and parks." A bit more: "Let me take you on a tour of my Houston where the azaleas have just finished a spectacular season and the roses are starting to bloom."People are jogging along the Bayou as we head downtown..." And finally: "We need to spread the real message about our great city. "Unfortunately it seems some of our messengers aren't up to the task".

The Buffalo Bayou deserves note. As Houston emerged from the homesteads of the early 1830s this tiny bayou was the artery of commerce, from the Gulf of Mexico/Galveston Bay right into what is now downtown Houston. Over the next 100 or so years, including World War II, it became the center of the petroleum refining and petrochemical industries of the United States. It was dredged extensively creating what is known as the Houston Ship Channel, bringing ocean going ships to within four miles of downtown. And it was polluted extensively. Environmentalists will argue that all of this was an unforgivable sin. Indeed if any of the polluters of this bayou were breaking the law at that time, then they surely should have been punished for that offense. However, the overall changes to the bayou, I would like to think, were part of the industrial growth process that has created the economic muscle of this country.

And with that success has now come a major drive to further clean up and improve this natural resource. East of downtown, San Jacinto State Park and Museum is a gem hidden amongst the distillation towers and oil tanks that prevail in that area. This park is a memorial to those who defeated Santa Ana in 1836. It also salutes the battleship USS Texas, and all who served on it, during World War I and World War II. Both the museum and the battleship are worth a visit. Suggestion: take Interstate 10 east of town, then the Lynchberg ferry exit (number 787) south. This route will avoid most of the refinery/petrochemical maze.

West of downtown this bayou is the home of several fine parks. First is Memorial Park, the *Central Park* of Houston. Others deserving mention include the Houston Arboretum, and a satellite facility of the Houston Museum of Fine Arts - The Bayou Bend Collection and Gardens. In addition there are several golf courses, plus an urban park along the Bayou, that runs into and through the downtown area. This area also includes a large number of grand residential areas such as River Oaks. Although vastly improved over the past 20 years, additional areas for rejuvenation of the Bayou remain[3]. Plans[4.1] include "building a contiguous system of trails on the north and south banks of Buffalo Bayou from Shepherd Drive to the Turning Basin" on the Houston Ship Channel. It soon will become as well known a resource to the Houston area, as the creeks and rivers of San Antonio, Dallas and Austin.

There are many other areas of addition or improvement to Houston area parks. Hermann Park[4.2], near the Medical Center and Rice University, has undergone a major upgrade, befitting its role, *as a best cousin* to Memorial Park. Next, Lake Houston Wilderness Park[4.3], part of the State Park System, is 30 minutes north and east of downtown Houston. Not far away is the Redstone Golf complex, home of the Houston Shell Open.

Finally I will close this litany with a bit on a rather eclectic neighborhood called the Houston Heights. This near downtown locale is so-named[5.1] because it is all of 23 feet above the downtown elevation. It is the home of many old Victorian style homes, many of which have had substantial rejuvenation. It is also the home of many brand new Victorian style homes, plus a polyglot of other dwellings. On balance it is site of considerable ambience, but zero views due to its elevation. Two trails deserve noting.

· First is a 1½ mile in length. It wanders[5.1] down the center of the esplanade on Heights Boulevard. Not unique, not without some problems, but on balance very nice.

· The second trail is much longer, say 5-10 miles. It is based[5.2] on land from old railroad tracks converted to the Houston Heights Hike and Bike Trail. This trail runs from the far NW corner of the Heights, south then SE towards I10 and the White Oaks Bayou to near downtown. Again it is not unique as many locations have replaced old rail tracks. But as the writer of the above site testified: "What a fabulous addition."

Perhaps urban scholar, Joel Kotkin, stated[6] it rather precisely: "if you want to see successful 21st-century urbanism, hop down to Houston and the Lone Star State."

•This is a Green City.

As one discusses environmental and conservation issues, the color of green is frequently introduced. Is one green enough? The answer to this question depends on how one defines environmentalism. I liked the original conservation of Theodore Roosevelt (TR). TR knew first hand about the need for protection of animals, for managing of grazing land and the nations forests. He also saluted the beauty and aesthetics of the west and worked to preserve the best. Color me dark green based on the original conservation of Theodore Roosevelt.

In contrast, particularly over the past 40 plus years, the environmental movement has moved beyond this macro approach to concentrate on the micro. This deals with molecules and particles, coupled with very low probability effects or very long term effects. It is the green of the Snail Darter, the green of the invisible, the green of the future long removed.

Now as one steeped in chemistry and climate change science one must deal in molecules or particles or in the far future, and that is not wrong. It is when these factors are extended to the extremes that concerns start to emerge. An example: demanding that nuclear waste disposal be adequate to last 10,000 years, perhaps even 100,000 years. This completely ignores the glacial cycles, and the very high odds that we will start to move into another devastating ice age over the next say 500 to 5000 years.

Micro environmentalism deals in subjects where the average citizen can never go. It deals in parts per billion or parts per trillion and the analytical chemistry to measure at these levels. Frequently the results—no matter how imprecise or uncertain—are shown on dazzling, multi colored maps, that convey high precision and certainty. Now, once again, there is nothing wrong with going to these places, or in using this technology, if it is done properly, objectively and kept in context. The problem is that the average citizen can not possibly understand all these technical subjects, and hence must rely on middlemen to define and explain the situation. The big risk here is the insertion of:

· government bureaucrats, who may come with specific agendas;

· Government regulators, who may be sincere, but not always sharp and

· *a new set of medicine men*, in command of the debate, answerable to no one, and able to dictate what can and what cannot be done.

Until this situation is rectified, until a sound method of accountability for these new kind of micro environmentalists is implemented, color me a very weak green.

• Section Organization.

There are two additional chapters in this section.

§7.2 will highlight the career of Theodore Roosevelt (TR). This story on Roosevelt covers his activity on conservation. TR can also be viewed as the founder of the macro environmental movement. While very much saluting him on most of this effort, the question is raised in this chapter if he ignored the *Law of Diminishing Returns?*

§7.3 will cover the career of Rachel Carson (RC). RC was the founder of the micro environmental movement. Her 1962 book, *Silent Spring*, became the flag pole for these efforts. While saluting her on some of her efforts, the question is raised if she also ignored the *Law of Diminishing Returns?*

•Conclusions.

If we set aside the comments from Jennifer Lazlo-Mizrahi, as that of a carping, political hack, one might ask if there is any substance to the charges that Houston had replaced Los Angeles as the most polluted city in the country. Well as it turns out, not very much. The basis for this claim is extremely thin, namely that

· for two years, and

· on only one out of six criteria pollutants — ozone — Houston barely exceeded LA. The reason for this *honor* was not any dramatic increase in the ozone levels or frequency of occurrences in Houston, but rather an amazing and commendable decline in these values for LA.

The above commentary is not to claim that there are no serious environmental problems left in Houston. There are, and people are working on them. The purpose here, however, is to emphasize that the political hacks have not *shot square* with the American public on this issue. They have deliberately insulted and scarred a rather fine city, and one that is doing more than it's share to solve the energy and environmental problems of our country.

And if the Left is not shooting square, on this issue—if they will use anything, say anything, and do anything to try to make their point—one has to wonder what are the other issues where they behave the same?

References and Notes.

(0) Poor, Jeff, *Iwo Jima Veterans Blast Time's 'Special Environmental Issues' Cover*, Business & Media Institute, April 17, 2008.

(1) Levin, Marc, *Democratic Spokesman Says "Houston is Filthy"*, Houston Review, September-October 2000. See: www.houstonreview.com/articles/FilthyHouston.html.

(2) Lanier, Elyse, *It would help if Houstonians stopped selling our city short*, Houston Chronicle, April 8, 2001.

(3) Snyder, Mike, *A better Bayou*, Houston Chronicle, September 22, 2002.

(4.1) See: http://www.buffalobayou.org/hikebike.html.

(4.2) See: http://www.houstontx.gov/parks/hermannpark.html.

(4.3) See: http://www.houstontx.gov/parks/lakehoustonpark.html.

(5.1) *White Oak Bayou Hike and Bike Trail*, as reported on Trails.com

(5.2) Fields, Mary Anne, *Houston Nature Walks*, houstonnaturewalksblogspot, October 13, 2006.

(5.3) Mullins, Cynthia, *Houston Heights Hike & Bike Trail*. See: http://blogs.har.com.

(6) Kotkin, Joel, *Houston: Model City*, Forbes Magazine, June 7, 2010.

7.2 On Theodore Roosevelt

• Background.

It is only fitting that I start this brief history of environmentalism with Theodore Roosevelt (TR), literally *the father* of conservation. I have seen his sculpture on Mount Rushmore. I have visited the national park in North Dakota named after him. I have seen the movie[0] *The Wind and the Lion* several times. And I have read several of his books. The book[1a, 1b] *The Rise of Theodore Roosevelt* was one of my favorites. This was a great story of a most unforgettable character. It also informed this writer that Roosevelt had something like150,000 letters to his credit, and was the author of fourteen books. This output is indicative of his capability, his energy, his many concerns, his desire to participate and his need to be heard.

• North Dakota Interlude.

TR, while a mainstream easterner, lived for parts of three years, from June 1884 to October 1886 in western North Dakota. This was triggered from the loss of his wife, who died of undiagnosed Bright's disease, and mother, from typhoid fever, both on February 14ᵗʰ, 1884. These deaths occurred two days after the birth of his first child, Alice.

I was struck by his courage, and his wisdom to get away from everything of his past life, at least for awhile. In June he headed to the Wildlands of North Dakota, on the far western side of that state. He was all of 25 years old. Here he learned how to be a cowboy, a rancher and a *rough rider*. I was impressed at his ability to handle the conditions he met, even thrive in them. I was impressed at his ability to interact with the people he met, even thrive on them. In short he completely fit in with his new environment.

He grew to love the beauty, vigor and challenges of the Dakota badlands and the frontier life. Clearly his life on the North Dakota frontier, and surrounding areas, is where he learned of the magnificent beauty and fragility of the west.

TR knew first hand about the need for conservation. He knew about the insanity of buffalo hunting, with the bison being slaughtered and nearly eliminated. He knew about game, in general, being severely over hunted and pastures being badly over grazed. Clearly conservation efforts were needed. He also saluted the beauty and aesthetics of the west and worked to preserve the best.

Roosevelt "has a claim on being the most interesting man ever to be president[2a]. He moved back east in 1886 to remarry, to resume his literary career, and to raise a large family. TR re-entered politics in 1889 rising to the governorship of New York in 1898. However his enthusiasm for reform was badly out of step with the power structure of the New York Republican party, who, as a result[3], "arranged for him to run as the party's vice presidential candidate in 1900. "When the republican ticket won, Roosevelt was out of their hair." However on September 14, 1901, President William McKinley was assassinated and TR became the 26ᵗʰ president. In 1904 he was reelected in a landslide. He served as president for 7½ years.

• His Presidency

TR has the reputation as a *trust buster* and indeed he initiated major anti-trust proceedings against the railroads, and many others, but most of his energy went into regulation. He sought[4] "to regulate, rather than dissolve, most trusts." He revived *The Sherman Anti-Trust Act of 1890* to restore needed competition in the world of business.

However, this will not be a major report on his presidency. Rather this effort will be limited to his position on, and his accomplishments in the area of conservation, along with some contemporary reflections. One quotation[5], at a *A Book-Lover's Holiday in the Open, 1916*, illustrates his interest, philosophy and priority on this issue. "Defenders of the short-sighted men—who in their greed and selfishness will, if permitted, rob our country of half its charm by their reckless extermination of all useful and beautiful wild things—sometimes seek to champion them by saying 'the game belongs to the people.' So it does; and not merely to the people now alive, but to the unborn people."

Included in his many contributions[6] were creation of several national parks and millions of acres of national forests and preserves. A summary of his efforts in this area is listed below:

· 150 National Forests (increased area from 43 M to 194 M acres);
· 3 National Parks;
· 20 National Monuments;
· 55 National Wildlife Refuges (51 Bird Reserves, 4 game preserves);
· 228 total.

Among his many contributions were the establishment of three national parks: Crater Lake in Oregon, Wind Cave in South Dakota, and Mesa Verde in Colorado. While he was not the founder of Yellowstone National Park, he did expand it. And he fought in vain to create a National Park in the Grand Canyon area. Instead, through the *1906 Act for the Preservation of American Antiquities*, he was able to create a National Monument in 1908 and add a National Game Preserve later in that year. It finally achieved National Park status in 1919.

TR used the 1891 *Forest Reserve Act* to set aside public lands as National Forests. Most importantly, he transferred the Forest Service from Interior to the Agriculture Department in 1905 based on the belief that tree husbandry or silvaculture was, in fact, farming.

TR sponsored 24 reclamation projects that stemmed from passage of *The Irrigation and the Newlands Reclamation Act of 1902*. This act led to the creation of the Bureau of Reclamation. One of these reclamation projects was the Salt River Project, east of Phoenix, which included a 260 foot high dam, that bears his name.

TR[7] also supported the *Homestead Act of 1862*, and "enabled two expansions during his tenure". TR wanted the land populated. "He and

147

his people loathed 'idle dreamers' who'd let the land sit indolent." As such he pushed for settlement.

His original environmentalism and original conservation efforts embraced the following concepts and principles.

· Included conservation of natural wonders: parks, monuments and wildlife refuges.

· Conservation is for the people as a whole. Results are not to be reserved for the few and the elite.

· Conservation includes development as much as it does protection, including water power, flood control and land reclamation. "Water[7] flowing unused to the sea was 'wasted'".

· Development means people and homesteads and dams and irrigation.

· Reclamation is[7] "claiming and harnessing. "Unused land or water needed to be reclaimed."

· Users of natural resources do not have the right to waste them, or to exhaust them.

· Governments, also do not have the right to waste them, or to exhaust them.

· We cannot reap without sowing. We cannot consume without husbanding.

· Corruption must be fought on many fronts including corrupt and unprincipled editors. One can extend that to modern times by adding corrupt and unprincipled environmental editors.

· Citizens will not be permitted to exterminate all useful and beautiful wild things.

One final comment on his presidency is in order. This was the successful conclusion of the Treaty of Portsmouth that ended the Russo-Japanese War in 1905. For his effort TR became the first American winner of the Nobel Peace Prize. His daughter, Alice—who was growing into a dashing and vivacious young lady—was part of the diplomatic party, and may have contributed[8a] to her father's success by her "ability to keep the press at bay by becoming the center of attention."

However, TR had a small problem with Alice, for Alice broke all the rules..Some have said she"drove her father crazy." And she may have ultimately passed her father in popularity and appeal. When once asked by a dignitary if he could better control[8b] his daughter he responded: "I can be president of the United States, or I can control Alice, I cannot possibly do both."

• Early Criticism.

TR surely did much good in all of these efforts. Yet some have accused him and his government of being highly autocratic in these efforts. Further, over 250 national items—parks, reserves, monuments and projects—is a huge number. An important question is did he go too far? Should some of these been left for state or local government, or for free enterprise?

TR was clearly a lover of *Big Government*. This may well have been a product of the times. Rampant corruption and huge trusts were the rule. I suspect the country, at that stage in it's development, required a big hand on the rudder of the ship of state. Somewhere along this pathway, however, he couldn't get off. He couldn't visualize the *Law of Diminishing Returns*.

If he ever heard of the idea that the federal government should limit itself to only a few crucial roles and do those exceptionally well, he never acknowledged it. He couldn't understand that he and his cronies were not always going to be right. He couldn't visualize, and perhaps did not understand that bureaucracies ultimately rise to their level of incompetence, to borrow a phrase from the *Peter Principle*. While TR was a highly motivated, involved and effective bureaucrat, he could not see that it would be impossible to achieve this standard across all the bureaucracies he founded, or to sustain this standard into the future. Nor did he see that the tools he promoted—the new laws and the new bureaucracies—would some day be *wasted* on political campaigns by unscrupulous politicians.

Finally TR did not have much experience with free enterprise. His only experience was his ranch in North Dakota which was ultimately sold at a loss[1c]. And even that might be hard to claim as true free enterprise. Perhaps this ranch was more of a badly needed escape mechanism, a way station in his life at that time. Perhaps it was also a way station for his travels about the west and even simply a hunting lodge. In reality, he did not need to make it a success. Surely he never ever shirked from his share of the work and more, and had little regard for a man who would.. A couple of quotes[8c] will illustrate this trait.

· "The first requisite of a good citizen of this Republic of ours is that he shall be able and willing to pull his weight."

· "No man needs sympathy because he has to work. Far and away the best prize that life offers is the chance to work hard at work worth doing."

While understanding these traits he never had the joy of starting from scratch and struggling for years to build a valuable entity. He never really experienced or knew the positive power of free enterprise.

In spite of his weaknesses and errors he still remains as one of this countries leading conservationists. Recently there has been some efforts—by[7] "'Range Writers', editorialists and 'educators'"—that claim Bill Clinton was a second coming of Theodore Roosevelt. The comparison is prompted, in part, via Clinton's activities on National Monuments. But one critic did not think much[9] of this comparison. This "administration unabashedly names new national monuments, closing land off to human activity, to appease its radical green supporters during every Presidential election year. "Always the designations have come despite the strenuous objections of residents, state and local government officials, and congressional delegations."

Perhaps the best squelch of this comparison can be seen in the following quote[7] from a TR speech: "There is not in all America a more dangerous trait than the deification of mere smartness unaccompanied by any sense of moral responsibility."

• Western Paintings Country.

Grandeur can best be experienced by first hand observation. However this is just not always possible on many of the grand natural settings. Books can be useful substitutes, and books devoted to paintings can help by depicting the classical grandeur of our planet.

For myself, I have a fondness for western paintings. The painter, Charles Russell (1864-1926) and his peer, Frederick Remington (1861-1909) captured the west. Both of these painters literally started their careers as *illustrators,* producing pictures that told a story. In his early years Russell would frequently include a detailed sketch—sometimes colored, on some western scene—in a letter to a friend. And Remington's first sales were to magazines. Theodore Roosevelt greatly admired the work of these painters. He was a peer in the sense of being a fellow lover of the west, and a prolific writer and fellow *story teller* of the west. Remington illustrated at least one of Roosevelt's books.

The *illustrating experience* of these two painters served them well, and led to a style that gave them the reputation that *no detail escaped their attention.* They produced several thousand paintings over their careers. Together they captured the grandeur of the west they knew and loved, as it slowly slipped away.

• More Recent Criticism: Progressivism.

TR, in addition to being the *Father of Conservation*, might also be thought of as the *Father of Progressivism*. He surely saw[10] his office very differently than his predecessors. TR believed in "presidential prerogative power" namely the ability to do anything not specifically prohibited by the Constitution or by any statue. Indeed "his first administration - - - forever changed the role of the President."

Perhaps no other individual has shone the spotlight on Progressivism more than Glenn Beck, the Fox News host and radio commentator. His interest in this subject in general and TR in particular led me to search the web, which led to an incredible odyssey of references. Some of these were rather strong tributes and some rather stinging critiques of TR. See NOTE 14 for details.

• Conclusions on TR

As noted above this was going to be some sort of odyssey. Clearly there are strong signs of Libertarian philosophy at work here, as well as other political views. And clearly this is a very complex subject as befits TR who was a very complex character. Perhaps it is too early to sort out the exact nature of TR's progressivism. This may be better left for future historians, particularly as this is part of a section on environmentalism and not on political philosophy.

References and Notes

(0) A review of the movie *The Wind and the Lion*, by a Michael Puttre on Amazon. com, captures my opinion perfectly. This movie stars Sean Connery, Candice Bergen, and Brian Keith. Connery is a Berber chief, Risuli the Magnificent, who kidnaps American Candice Bergen and her two children in Morocco in 1904. Connery wants to force the European powers out of Morocco, which at that time is occupied by the British, French, and Germans, each with different agendas. Brain Keith is President Teddy Roosevelt who sends the Marines to Morocco "to get respect." In the end, the Americans and Connery's Berbers make common cause against the Germans.

For both the reviewer and myself, Brian Keith, as TR, is worth the price of admission. Of particular fit to this chapter is the scene where he had to kill a huge Grizzly, or be killed. Then he had this bear mounted in a vertical position—to give it all the grandeur it deserves—and to display it prominently in Washington.

(1a) Morris, Edmund, *The Rise of T. Roosevelt*, Coward, McCann & Geoghagen, New York, 1979.

(1b) Roosevelt, Theodore, *Hunting Trips of a Ranchman*, G. P. Putnam and sons, New York, 1885. See also: www.Bartelby.com, 1999.

(1c) The winter of 1886/87, was very severe, killing cattle by the tens of thousands. Although Roosevelt missed this winter, barely, it gives a feel for how tough a Dakota winter could be.

(2a) Morris, Edmund, *Theodore Rex*, Random House Inc., New York, 2001. This particular quote comes from Robert Kirsch of the Los Angeles Times Book Review, as reported on the book jacket for this book.

(2b) Ibid

(2c) Ibid

(3) PBS, New Perspectives on the West: *Theodore Roosevelt (1858-1919)*, 2001. See: www.pbs.org/weta/thewest/people/i_r/roosevelt.htm

(4) Encyclopedia Americana, *Grolier presents The American Presidency: Theodore Rooseveelt*, © 2000 Grolier Incorporated. See: http://gi.grolier.com/presidents/ea/bios/26proos.html.

(5) Theodore Roosevelt Association, *Quotations of Theodore Roosevelt*, undated. See www.theodoreroosevelt.org/life/quotes.htm.

(6) Theodore Roosevelt Association, *Conservationist Theodore Roosevelt*, undated. See www.theodoreroosevelt.org/life/conservation.htm.

(7) Brossman, Dennis, *Which Teddy do you mean*, eco-logic on-line, February 1, 2001.

(8a) Alice Roosevelt Longworth, from Wikipedia.

(8b) Theodore Roosevelt, from Wikiquote.

(8c) *Some Theodore Roosevelt Quote, by Topic*, undated. See: http://users.metro2000.net/~stabbot/trquotes.htm

(9) Randall, Tom, *Uncle Sam: Sell that land!*, Environmental & Climate News, March 2000. See also: www.heartland.org/environment/mar00/editorial.htm.

(10) Streich, Michael, *Progressivism and Teddy Roosevelt - TR's First Term in Office Produces Significant Victories*. See: http://modern-as-history.suite101.com/article.cfm/progressivism-and-teddy-roosevelt.

(11.1 - 11.6) See NOTE 14.

7.3 Rachel Carson

• The Intervening Years

Interest in conservation did not end when TR left office. Indeed a recent time-line—that lists[0] "the many events that occurred before the publication of Rachel Carson's *Silent Spring*"— gives an indication of such interest. However, over the interval of 1914-1962—what with World War I (WWI), then the depression and dust bowl, followed by WWII, the Korean War, the and the Vietnam War—most governments did not have much spare time or other resources to allocate to the environment. During this period, particularly after WWII, there were many environmental incidents and disasters, a few which are noted below. Many of these disasters were deadly incidents of air pollution. Several incidents cited were for water pollution via toxic metals.

Clearly action was needed, and California led the way with the establishment, in 1947, of the Los Angeles Air Pollution Control District, the first in the nation. The London incidents ultimately led to the British Parliament passing their Clean Air Act in 1956. The Donora incident contributed to passage of the Federal Pollution Control Act of 1955—a forerunner of the Clean Air Act of 1963—which focused on research into the causes and effects of air pollution. And in 1959 California became the first to impose automotive emissions standards. The Japanese incidents contributed to 14 pollution control laws, including amendments to existing laws, being passed by the Japanese Diet in 1970. This Diet was known as "The Environmental Pollution Diet".

Site	Year or Years	Event	Casualties	Causes
Japan, Minimata Bay	53-59	Minimata disease	67 dead, 330 hurt	Mercury poisoning via eating fish.
Japan, Toyama	53-59	Itai-Itai disease	-	Cadmium mine leaks.
UK, London	1948	Killer fog	600 dead	Coal heating
UK, London	1952	Killer fog	~ 4,000 dead	Coal heating
UK, London	1956	Killer fog	1,000 dead	Coal heating
USA, Donora, PA	1948	Yellow fog	20 dead, 6,000 ill.	Blast furnaces, zinc emissions, trapped air, high SO2 levels.
USA, New York, NY	1953	Smog incident	170-260	Inversion
USA, New York, NY	1963	Smog incident	405	Inversion
USA, New York, NY	1966	Smog incident	168	Inversion

A Sampling of Critical Environmental Incidents: 1948-1966

• Her Early Years

It is only fitting that I end this brief history of the embryonic days of environmental concerns with Rachel Carson (RC). She surely was the founder of the modern micro and sub-micro environmental concerns through her 1962 blockbuster book[11a] *Silent Spring*.

Born in 1907, she was reported to have wandered the banks of the Allegheny[1b] with her mother, in what was then "the pristine village of Springdale, PA, north of Pittsburgh". As she grew up she watched as Springdale[1 c] was "transformed into a grimy wasteland, its air fouled by chemical emissions, its river polluted by industrial waste. "She observed that industry took no notice of the defilement and no responsibility for it. "The experience made her forever suspicious of industry.

Carson started college in 1925, as an English major, but switched to biology. She later earned a summer fellowship at the Woods Hole Marine Laboratory, where her fascination with the sea was initiated. She received a scholarship to John Hopkins, completing an MA in zoology, in 1932.

She supported herself, as a part time lab assistant and by free-lance writing. Some of this was for the Baltimore Sun on the nature and history of Chesapeake Bay. In 1936 she landed a job as a writer for what soon would become the U. S. Fish and Wildlife Service (USFWS) . A year later she became a junior aquatic biologist for the USFWS. She was promoted to aquatic biologist in 1943.

RC undoubtedly was restricted in her scientific work because of her education and her gender. However her talents in writing were immediately recognized and she was assigned to edit the field reports prepared by other scientists. This provided her with an in-depth insight into all the research underway in this agency. By 1949 RC was named chief editor for all of their publications.

Year	Book	Comments
1941	Under the Sea-Wind	-
1951	The Sea Around Us	A best seller for over a year, and serialized in the New Yorker. Won the National Book Award.
1955	The Edge of the Sea	A best seller.
1962	Silent Spring	A best seller for a year.
1965	The Sense of Wonder	Published posthumously.

Rachel Carson - the Author.

She clearly liked to write, and ultimately wrote several books. She was able to retire from the USFWS in 1952. But she did not retire from further writing. With the release of *The Edge of the Sea*, Carson "became the foremost science writer in America".

RC planned[2a] to write additional books about "the science of ecology and the intricate relationships that govern the natural world". These plans[2b] were, however, interrupted by the emergence of new "pesticides that had been embraced by American farmers" at the end of WWII. "Hundreds of thousands of acres of forest and crop land were sprayed." Evidence started to surface on descriptions of severe damage to wildlife and to the emergence of soft egg-shells for some birds. Carson began to investigate this evidence and, in essence, this became her lifetime mission. Her writings with Silent Spring "told a terrifying story about the effects of chemical pesticides - - - ". However the initial chemical[2c] "that Carson mentions in *Silent Spring* is not DDT but the radioactive isotope Strontium 90. "Throughout the book Carson draws upon the public's familiarity with deadly radioactive fallout - - -." She compares this isotope with DDT, but. according to a 1962 book review [3] in the New York Times, she "fears the insidious poisons spread as sprays and dust or put in foods, far more than the radioactive debris from a nuclear war."

• **The Impact of Silent Spring**.

This book was both widely praised and condemned. It was surely not ignored. It led to a vigorous and heated debate ultimately leading to the ban of DDT.

(1) This book was widely praised. While praising RC is not the role of this report, there is much about Rachel Carson with which I can salute. She was surely a lover of nature and could see that society was not doing right by nature. I would have little trouble saluting the aspects of her career that demonstrated insight, courage and tenacity. She was also something of a maverick, with no official scientific laboratory or governmental agency to support her views over her last 12 years. As a self proclaimed maverick, as noted at the start of this book, I would have great empathy on what such a pathway entails. I would also have little trouble supporting her original macro environmental views:

· as she watched her hometown of Springdale being destroyed;

· as she saw the near destruction of Chesapeake Bay by industrial drainage;

· as she learned about ocean pollution in Japan, specifically on Mercury and Cadmium;

· as she found out that the pesticide industry did not do adequate testing and did not provide appropriate application instructions. Thalidomide, while not an insecticide, was an example of this problem and what can happen. It was another female scientist, Frances Kelsey, at the U. S. Food and Drug Administration (FDA), who kept this drug from being released in the States[12d].

(2) This book was also widely condemned. The critics attacked her capability, credentials, credibility and character rather mercilessly, in spite of her accomplishments as an author and as a women who demonstrated over and over again that she had much grit[4]. Many of these attacks were efforts at intimidation. The agricultural chemical industry[1e] "was not about to let a former government editor, a female scientist without a PhD or an institutional affiliation, known only for her lyrical books on the sea,

to undermine the public confidence in their products or to question its integrity". They called her a "bird and bunny lover". She was even accused of being a communist.

However more recent and objective critiques of her work have been written as noted below.

A Sampling of Key Critiques on Rachel Carson and Her Position on DDT

· Dr J. Gordon Edwards taught entomology/biology at San Jose State U. for 43 years. A member of the Sierra Club and the Audubon Society, he is a fellow of the California Academy of Sciences. Dr Edwards prepared this report[6], *The Lies of Rachel Carson*, in 1962-63, but did not publish it until 1992. This is a very damming essay. He charges Carson "was not interested in the truth about these topics, and that I really was being duped - - ." In the first chapters I saw "many statements that I realized were false." Near the middle the feeling grew that RC was being "loose with the facts and was also deliberately wording many sentences in such a way as to make them imply certain things."

· Dr. Steven Milloy and Dr J. Gordon Edwards. Milloy is an adjunct scholar at the Cato Institute, a writer for FoxNews.com and the sponsor of junkscience.com. He has 4 degrees including a Juris Doctorate and a Master of Health Sciences in Bio-statistics. Edwards & Milloy prepared[7] a list entitled *100 things you should know about DDT.* The inputs are primarily from the materials compiled by Edwards. In this list the writers charge that "DDT was demagoged out of use." The 109 entries are arranged as: I. History; II. Advocacy vs DDT; III. EPA hearings; IV, V. Human exposure, cancer; VI - X. Birds: egg shell thinning, eagles, falcons, bird numbers; XI. Erroneous detection.

· Gordon S. Jones is a writer and political scientist, and has spent 30 years in Washington, with 30 months at the EPA, administering the new ban on DDT. In an essay, in the 08/26/02 issue of Human Events, entitled *The Destructive Legacy of Rachel Carlson*, Jones notes that "DDT was banned as a result of Rachel Carson's literary skills." The "ban made no sense to me then and it makes no sense to me now". This ban was "responsible for 60 million deaths." Yet DDT has saved over 100 million.

· Dr. Gilbert Ross is the Medical Director of the American Council on Science and Health (ACSH) in New York.. See: sepp.org/NewSEPP/DDTban.htm for his essay titled *Banning DDT Does Not Make Sense.*

This essay is based on inputs from July -September 2000 issues of The Lancet[12b]. After WWII, DDT stopped a typhus epidemic. It was next used/overused as an ag. pesticide. DDT remains the most effective tool against malaria, sleeping sickness & yellow fever.

· Michelle Malkin is the daughter of Filipino immigrants. She is an editorial writer, and co-author of the report[13] *Rachel's Folly - The End of Chlorine*. In an 10/13/00 editorial in the Washington Times, Malkin goes after the anti-pesticide patrol. This Patrol "claims to work for the most vulnerable members of society. "But it is the children and the elderly who are most at risk of illnesses that will spread if mosquitoes are not controlled."

Several additional inputs from the British medical journal The Lancet are also noted below.

Selected References on DDT from The Lancet

· Editor. An editorial in Lancet, on 07.22.00, *Caution required with the PP.* The Editor noted, that proponents of DDT, have used the Precautionary Principle {PP} to push for a global ban on DDT. Concludes the health of people in poor countries is being put at a very real risk to protect the citizens of wealthier nations from a theoretical risk.

· A. G. Smith, MRC Toxicology Unit, Leicester University. In the 07.22.00 issue, Smith asked: *How Toxic is DDT?* This is an excellent overview of DDT impact on humans. "There are probably few other chemicals that have been studied in as much depth as has DDT, experimentaly or in human beings." Dermal and oral toxicity, carcinogenicity and reproduction function show very little if any impact from DDT expossure.

· D. R. Roberts et al, Uniformed Services, University of the Health Sciences. In the 07.22.00 issue, this paper, *DDT house spraying*, reported: "Malaria is reappearing in urban areas." - - "Although many factors contribute to increases, the strongest correlation is with lower number of houses sprayed with DDT."

· G. Ross, American Council on Science and Health. Finally, in the 09.30.00 issue, wrotr: *The DDT Question.* Prior inputs: "Makes abundantly clear the profoundly harmful effects of the - - banning of DDT - -. "The real tragedy is that there is not - - any valid reason for these measures, which have caused so much human suffering. "They are the unfortunate legacy of the self-styled environmentalists, whose bible remains RC's Silent Spring."

(3) <u>This book triggered legislation against DDT</u>. Her work was brought to the attention of John F. Kennedy, and in early 1963 pesticide policy development was shifted[15] from the Department of Agriculture to the Department of the Interior. However, RC did not live to see all of the impact of her work. She must have gained great satisfaction over the reception of her books, and the recognition that her work would be carried forward. Eighteen months after the release of Silent Spring she passed away, at the age of 56. Six years after her death the National Environmental Policy Act was passed, establishing the Environmental Protection Agency. And in 1972 the production of DDT was banned in the United States. "She had set in motion a course of events that would result in a ban on the domestic production of DDT and the creation of a grass-roots movement demanding protection of the environment through state and federal regulation".

• A critique of Carson's Writings.

There surely is no argument that Carson's work, and her presentation of it, have had a major impact. But there is an argument if her presentation was intellectually and actually honest. What follows are inputs on her biases, on the slanting of her writings and the possibility that her writings contained many falsehoods.

(1) Biases against other scientists, bureaucrats and businessmen.:

· Was she suspicious of the objectivity of many male government scientists as a result of the chauvinistic treatment she received while in the government?

· Did she have little confidence in the wisdom of many government bureaucrats because of their decisions on pesticides and because of what she would conclude was their lack of leadership while she worked at the USFWS?

· Was she so poisoned by her experience with industry in Springdale, that she could not have any trust whatsoever with businessmen?

I believe the answer to each of these questions would be yes.

(2) <u>Slanted use of her high horsepower writing skills</u>.

· Did she use her colorful, poetic, hypnotic and almost terrifying writing style to help her win the day? Absolutely. In the first chapter in *Silent Spring*, RC talks[1f] of a town and surrounding countryside of great beauty and harmony with nature. Then she gradually transforms this town into stillness and death. "Even the streams were now lifeless. Anglers no longer visited them, for all the fish had died. "In the gutters under the eaves and between the shingles of the roofs, a white granular powder still showed a few patches; some weeks before it had fallen like snow upon the roofs and lawns, the fields and streams." RC admitted that no such town exists. She may very well have been using what the steel, coal and coal-tar industries did to Springdale to make her point. She indicates that she couldn't wait to get out of Springdale because of pollution. I have seen coal dust spread over ones window sills and ones car. Hence I can relate to the moving if such a problem is not resolved.

· Was she a fear-monger? Lear reports that some[2e] regard "*Silent Spring* as one of the finest examples of apocalyptic literature - - -". She also reports[2c], to other teachers, that the very first chapter in *Silent Spring* may provoke discussion "about the value of apocalyptic writing as a tool of social reform". In another reference[3], the writers note that she "intends to shock - - -".

· Did she tilt her presentations to assure victory? Absolutely. For example the Chemical and Engineering News magazine commented [3] on the articles published in The New Yorker as follows: "The articles are obviously, perhaps also intentionally, one sided". In another example, Lear reports[12e] that other "recent chroniclers criticize Carson for failing to include a more bio-centric approach to her critique of pesticides". Finally the Milnes[3] comment that "'Silent Spring' is so one sided that it encourages argument - - -".

Clearly I believe the answer to all of the above questions would be yes

• **Unethical use of her high horsepower writing skills.**

Did Carson go beyond slanting her claims, by lying to the American public and to the world? Well, Dr. Edwards[6] surely believed she lied as noted earlier. She "was playing loose with the facts." Further she was lying by omission "omitting everything that failed to support her thesis that pesticides were bad, that industry was bad, and that any scientists who did not support her views were bad". In testimony and in speeches Edwards would present statements from Silent Spring and then read "the truth from the actual publications she was purporting to characterize. "This revealed to the audiences just how untruthful and misleading the allegations of Silent spring really were."

· Carson wrote extensively about egg shells thinning. Edwards and Milloy[7] provide 26 citations that either refute this claim, or indicate other factors would be the cause of any observed thinning. One example[18]: "Among brown pelican egg shells examined there was no correlation between DDT residue and shell thickness".

· In a related area[6] , Carson cited James DeWitt of the USFWS on quail/pheasant experiments. For example the quail were fed 100 ppm of DDT, in all their food, every day for an extended period. (3000 times the daily intake humans would have received during the years of maximum DDT application). Carson reported that this experiment proved that the quails reproduction was seriously impacted. While the quail egg production remained normal Carson claimed "few of the eggs hatched". However DeWitt had reported that 75 to 80 percent of the quail eggs hatched. This can be compared to the controls—those quail that received no DDT—where 83.9 % of their eggs hatched.

· Carson wrote extensively about birds dying off and frequently referred to this extinction. She "alleges that because of the spray programs 'Heavy mortality has occurred among about 90 species of birds'." She provides no references to confirm these charges, and completely ignored bird counts, such as The Audubon Christmas Bird Counts[6] , that reveal a totally different picture. There is no way that such a bird lover as RC would not know about such counts.

· Another lie reported by Edwards notes[6] that Carson claimed the FDA "forbids the presence of insecticide residues in milk shipped in interstate commerce". The allowed level is 0.5 ppm. Now this sounds like such a small amount, but it can be restated, to borrow a trick from the environmentalists, as 500 ppb. More importantly there is no way that RC would not know about this limit.

· Carson also wrote extensively about cancer. Lear reported[1g] that, in "one of the most controversial parts of her book, Carson presented evidence that human cancers were linked to pesticide exposure". Edwards and Milloy[17] noted that "DDT was alleged to be a liver carcinogen in *Silent Spring* and a breast carcinogen in *Our Stolen Future*[9]. They cited thirteen references to refute this claim. One example[20] : "Men who voluntarily ingested 35 mgs of DDT daily for nearly two years were carefully examined for years and 'developed no adverse effects'". There is no way that RC would not know about this report in the Journal of American Medical Association. Finally Gina Kolata, a science writer for the New York Times, has written extensively on this issue. One example[11] : *Study Discounts DDT Role in Breast Cancer.*

I believe the answer to the questions—did Carson lie directly and/or did she use the cleverly designed lie—would be yes and yes. RC lied to the

American public and to the world. Carson used all of her writing skills to make her case, no matter what the cost. Here, the cost was her honesty.

• A Critique of Carson's Judgement.

There surely is no argument that Carson was a great author. Beyond her writings, questions emerge as to her reasoning and analytic skills and her overall judgement. I will discuss two areas.

(1) <u>Her disdain for the *Law of Diminishing Returns*</u>. Did she, like Theodore Roosevelt, ignore the law of diminishing returns? I would also believe yes on this question.

· For example, Lear notes[1g] that Carson "categorically rejected the notion proposed by industry that there were human 'thresholds' for such poisons - - - ". Zero tolerance was her position. Not 1 ppm, not 1 ppb , not 1 ppt, but zero.

· Another example. Carson called all insecticides *agents of death* While this maybe true in the sense that they killed insects, her implication was much broader than that. She frequently would couple insecticide development to chemical warfare, but this couple was greatly exaggerated by Carson, and frequently her time-line was completely off. Indeed as one studies the citations from the Lancet[12b], one comes away with a totally different picture on the value of insecticides. They surely were not all *agents of death*, in the broader context that she used.

· A third example. Carson seemed to categorically reject all chlorinated chemicals. While talking about the chemical chlordane, she makes a blanket indictment[1h], that, "like all chlorinated hydrocarbons its deposits buildup in the body in a cumulative fashion". But chlordane is not like all chlorinated hydrocarbons. It is one of the most complicated compounds ever made. This molecule contains an almost unheard of eight atoms of chlorine. In contrast, DDT contains five chlorine atoms, dioxin has four and the simplest chlorinated hydrocarbon (methyl chloride) only a single chlorine atom. It would seem to Carson that if a molecule with 8 chlorine atoms is very bad, than one with 5 chlorine atoms must also be very bad, and if a molecule with 5 chlorine atoms is very bad, than one with 4 chlorine atoms must also be very bad, and so on down to even a single atom. Indeed some environmentalists[23] are even going after chlorine charging, for example, that its use in water purification is totally unsafe.

While this writer is convinced she went too far in many areas, some of her supporters even criticized [12e] her for "not going far enough in challenging big business/scientific establishment".

(2) <u>Her disdain for the *Art of the Compromise*</u>. Much of the above section could also fit here, such as her position on zero tolerance. But by far her most amazing lack of compromise is between the non-use of DDT and it's use against malaria. The most amazing sin of omission in *Silent Spring* is the total lack of attention to malaria, with only a single entry in the index. This neglect is almost unforgivable. Edwards commented[6]: "Surely Carson was aware that the greatest threats to humans are diseases such as malaria, typhus, yellow fever - - - . "She deliberately avoids mentioning any of these, because they could be controlled only by the appropriate use of insecticides, especially DDT."

Did she mislead the United States, and indeed the whole world on the use of DDT? I would also believe yes, based on the inputs highlighted in the critiques covered in this chapter. The statistics, associated with malaria are staggering. Four authorities in this table will be noted:

· Gordon Jones reports that DDT is credited with saving 100 million lives, while the ban on DDT was responsible for causing 60 million deaths.

· Edwards and Milloy in their 1999 report [7] "To only a few chemicals does man owe as great a debt as to DDT. DDT has prevented 500 million human deaths, due to malaria, that otherwise would have been inevitable." This citation was from the National Academy of Sciences [24]

· Dr. Terrence Kealey, vice-chancellor of Buckingham university noted in 2001, "Worldwide, malaria has returned with a vengeance, accounting annually for 300 million cases, and, sadly, one million deaths, mainly of children". There is no indication that she was aware that there is a great moral issue here. Lear was aware of this issue[12e] noting that some criticize Carson "for neglecting the issues of social justice and equity for the decisions of who sprays and who gets sprayed..

• **The Ban on DDT and the Impact on Malaria Control.**

Details on this complex and controversial subject can be found in NOTE 15.

References and Notes

(1) Neuzil Mark and Kovarik, Bill, Green Crusades: Environmental Conflict in History. See:
www.geocities.com/combusem/HIST.HTM. Provides inputs from ancient times up to *Silent Spring* in 1962. See also: Kovarik, Bill, *Environmental History Timeline: 1940 - 1960*.
See: www.radford.edu/~wkovarik/hist1/7forties.html.

(1a) Carson, Rachel, Silent spring, Houghton Miffin Company (2002), Boston. This is the 40th Anniversary Edition, with a new introduction by Linda Lear. This introduction is available at the Barnes & Noble web site for this book. See also Table 6.4 for other books by Rachel Carson.

(1b) Ibid, page xi.

(1c) Ibid, page xiii.

(1d) Ibid, page 211.

(1e) Ibid, page xvii.

(1f) Ibid, page 3.

(1g) Ibid, page xvi.

(1h) Ibid, page 24.

(2a) Lear, Linda Lear, *Rachel Carson and the Awakening of the Environmental Consciousness,*
Part 1, National Humanities Center.
See: www.nhc.rtp.nc.us:8080/tserve/nattrans/ntwilderness/essays/carson.htm.

(2b) Ibid, Part 2

(2c) Ibid Part 4. Both Part 4 and Part 5 have a sub-title: Guiding Student Discussion. Lear uses these two parts to give instructions or ideas to other teachers.

(2d) Ibid, Part 5. Dr. Kelsey received the Distinguished Federal Civilian Service Award in 1962 for blocking the release of Thalidomide into the United States. Kelsey's career was equally unique and important as Rachel Carsons.

(2e) Ibid, Part 6. Part 6 has a new sub-title: Scholars Debate.

(3) Milne, Lorus and Margery, *There's Poison All Around Us*, New York Times Book Reviews, September 23, 1962. See: www.nytimes.com/books/97/10/05/reviews/carson-spring.html.

(4) Her father passed away in 1935, when RC was all of 28 years old. Her sister died in 1936 and Carson and her mother raised her two orphaned children. Ultimately Carson looked after her mother. PBS reported that despite "innumerable personal tragedies while she was working on Silent Spring - - she was seriously ill, a niece died and left a young son whom she adopted, her mother died, and she learned she had cancer - - Carson produced a book that would take on a life of it's own". See: www.pbs.org/wgbh/aso/databank/entries/btcars.html.

(5) For example see *Notes from Rachel Carson's Silent Spring*, items 57 to 60. These give a flavor of the political stresses between the Agriculture and Interior Departments. These notes are at the email address for a Mark Barrow at Virginia Tech.
See: www.majbill.vt.edu/history/barrow/hist3706/readings.html.

(6) Edwards, J. Gordon, *The lies of Rachel Carson*, eco-logic on-line, November 1, 2002. See note 1.4 on instructions on how to gain access to this database.

(7) Edwards, J. Gordon and Milloy, Steven, *100 things you should know about DDT*, ©1999. See:
www.junkscience.com/ddtfaq.htm.

(8) Switzer, B., 1972. Consolidated EPA Hearings, Transcript pp. 8212-8336. Also see Hazeltine, W. E., 1972. *Why pelican eggshells are thin*, Nature, 239, pp 410 - 412. See note 44 in (6.17)

(9) Colburn, Theo, et al, *Our Stolen Future: Our We Threatening Our Fertility, Intelligence and Survival. A Scientific Detective Story*, Dutton Publishing, 1996.

(10) Hayes, W., 1956. Journal of the American Medical Association (JAMA) 162: 890 - 897.

(11) Kolata, Gina, *Study Discounts DDT Role in Breast Cancer*, New York Times, October 30, 1997. See also *Study: Exposure to DDT Doesn't Increase Risk of Breast Cancer*, at www.junkscience.com/news/kolata2.html.

(12a) See an above table for selected references on key critics of Silent Spring.

(12b) The Lancet is Britain's most prestigious medical journal. See an above table for selected references from The Lancet on DDT, form 1972 and 2000 issues.

(13) Malkin, Michelle and Fumento, Michael, *Rachel's Folly - The End of Chlorine*, Competitive Enterprise Institute, 1996. Malkin is also the author of the best selling book *Invasion*, on our nightmare immigration policies and their administration.

(14) See report by the NAS Committee on Research in the Life Sciences of the Committee on Science and Public Policy, 1970. The Life Sciences; Recent Progress and Application to Human Affairs; The World of Biological Research; Requirements for the Future.

(15 - 20) See NOTE 15.

8. Conclusions

8.1 Why I am a Skeptic on the Global Warming Issue

• **Background.**

Starting in §1 of this book I have painted a picture of the huge propaganda campaign underway in our country. I have defined Albert Gore Jr. as the chief male propagandist. In turn, I have defined Ms. Nancy Pelosi as the chief female propagandist. I have also spent some time on her quest for a coronation. In any event she is a very active propagandist. And finally I have covered a very broad picture of the energy, environmental and climate change scenes and ultimately answered the question as to how green are the Gorons?.

On the global warming issue, count me on the skeptics side[1]. While the number of skeptics has been reported as small, one petition[2] has been signed by about 20,000 scientists, engineers and others. As noted in Chapter 3.1, most skeptics are well informed with some coming across as positive, brilliant, and human, far from the mediocre scientists that they are frequently called.

Skeptics have been ridiculed. Such attacks are inconsistent with the National Academy of Sciences code of conduct, which indicates that "the fallibility of methods is a valuable reminder of the importance of skepticism in science. Scientific knowledge ... must be continually scrutinized for possible errors." The code goes on: "a searching skepticism as well as an openness to new ideas are essential to guard against intrusion of dogma or collective bias into scientific results." Unfortunately we are very close to this condition today.

• **Key Assumptions on Global Warming.**

There are four key assumptions on the global warming issue.
(1) Our planet is warming
(2) This warming is caused by society.
(3) This warming will be catastrophic.
(4) Society knows what to do to prevent this warming.

These assumptions were discussed in depth in Chapter 14 of my first book. A summary of that input is provided in NOTE 16.

• Comments.

All of the four assumptions have been assessed and found to be questionable. None of them can be said to have been confirmed to date. It is clear that the science behind global warming is complex and incomplete. But the consequences of starting down the pathway are so enormous that it benefits us all to make sure we get the science right and not settle for a politically defined solution, which is what the KP is all about. The KP would cost the North American taxpayer trillions of dollars, dramatically affecting our economies, with no assurance this investment would solve any problem.

It is my conviction that the proponents of this issue have not proven carbon dioxide guilty. Even more so they have not made the case that it is imperative we act immediately.

I do not see the catastrophic future painted by the proponents of the KP. It is far easier to see serious economic hard times if society constrains itself any further than it already has on nuclear power, that it has considered doing on hydro power, and that it would do on fossil fuels if caps were placed on carbon dioxide emissions.

Hence I do not support the KP which I believe is terribly flawed and possibly even fraudulent.

References and Notes

(1) Earlier versions of this essay were first published in:
* The Houston Chronicle, June 17, 2001, with Michael T. Halbouty, and
* The Austin Review, November 6, 2001, with Michael T. Halbouty, and
* The Green & White, University of Saskatchewan Alumni Association, Fall 2002..
(2) See Chapter 10, Note 6c, in my first book.

8.2 How Green are the Gorons?

A Variety of Inputs on this Key Question.

• On an Early Embrace of the Color Green

As noted at the start of this book—in the Dedication—my love for nature stems from my father. First of all he was a master gardener. Gladiolas were his specialty, but he also took great care with his Dahlias, Peonies and tomato plants. While I have never been a gardener, I grew up in a house full of gladiolas and peonies, and a love for fresh tomatoes. And I have frequently expressed my awe at the grandeur of the botanical bounty of Earth. For example in Texas: the Bluebonnets and the Indian Paint Brush wild-flowers; the Azaleas and the Hibiscus and their panorama of color; the Bougainvillea and the Crepe Myrtles, and their incredible long season in bloom; the Live Oaks and their magnificent profile and the Pin Oaks and their great leaf.

Indeed, I have been in awe of the color *green* for more than 60 years. This started in my high school days with my interest in science fiction. One of the most memorable inputs from that era was a poem from the short story *The Green Hills of Earth*[1] by Robert Heinlein. This short story is considered one of the true classics of science fiction.

An extract from that story—a brief part of a poem by the same name—is shown below. I can recall as a youth, reading this poem. It told a story about hypothetical space travelers and their yearnings to see once again, the cool, green hills of Earth.

The Green Hills of Earth in part
Robert A. Heinlein

We pray for one last landing
On the globe that gave us birth;
Let us rest our eyes on the friendly skies
And the cool, green hills of Earth.

Other artists have written or sung about this phenomena. One of these that expresses somewhat the same sentiment[2] as the wandering space traveler, would be the young, but dying cowboy in the song *The Streets of Laredo.*

The Streets of Laredo
as sung by Johnny Cash

Beat the drum slowly and play the fife lowly
Play the dead march as you carry me along
Take me to the green valley lay the sod o'er me
For I'm a young cowboy and I know I've done wrong.

Finally we have the story[3] of the lonesome frog, Kermit, who reminds one and all that *it is not easy being green.*

The above items are included to demonstrate that *green concepts* were embraced by this writer, long before environmental forces made it popular to be green.

• On the Spectrum of the Color Green.

Throughout this section, and in other parts of this book the color green has been noted. One of the more annoying aspects of the Gorons is their attempted kid-napping of the *Color Green.* They claim to be green, but surely this must be limited to a very narrow piece of the green color spectrum. A very special shade of green is needed for this lot.

The Wall Street Journal may have been the first business paper to talk[4.1] of *Shades of Green.*

This clipping reported that a *"Poll Finds Many Are Torn By Wish For Convenience In Their Buying Decisions."*

Peter Huber received an engineering education from MIT. In his 1999 book[4.2]: *Hard Green - Saving the Environment from the Environmentalists,* Huber may have been the first scientist to suggest there is more than one shade of green. For more on Huber see §6.6 on emissions control improvements.

And William Ruckelshaus (WR), in his essay[4.3]: *A New Shade of Green,* may have been the first environmentalist to suggest there is more than one shade of green. Here WR, of DDT fame, sounds almost like the second coming of Rachael Carson. For more on WR see §7.3 and his comments there on DDT.

Now the environmentalists would probably not like "soft green" as their color, the opposite to "hard green." They might like dark green, but I could not grant them that color, as I can recall a Camaro, in the British Green version of dark green, as one of my favorite colors.

Same reaction for Kelly Green and a true Lime Green. Now a sickly lime green might be a fit, for at times, the extreme environmentalists have become so extreme that one must wonder if they are coming down with something. In any event we must find a unique and distinctive color—this will be only a tiny piece of the green spectrum—to be reserved for them.

Of course there are some environmentalists/politicians that are more red than green. These are the *watermelons* of the world: green on the outside, but red on the inside.

And then there are some *Greens* that embrace the environmental red light: no one to the left, no one to the right, no one coming at me and no one behind me; hence the light is really green

A rather intriguing and original headline stated[5a] that: *Green is for Sissies.* This introduced a fairly balanced, 2008 situation review, on ExxonMobil, which leads me to suspect this headline was due to a New York Times headline writer, and not the oil company.

A second headline by the Times was almost as intriguing. It stated[5b] that we live in a land of *Green Guilt* where even the most committed confess that they commit environmental sins. This input surely has a religious overtone to it.

Indeed several writers have commented that environmentalism is close to a religious belief system. A rather unique perspective[6] on such an analogy notes that "it provides its adherents with an identity," and as "the world becomes less religious, people can define themselves as being Green rather than being Christian or Jewish."

• Some Comments on Gwyneth Paltrow.

A commentator came up with the key question: how green is Gwyneth Paltrow? A peak at her movie *Great Expectations* give hints of many shades of green such as: very light, celery, lime, pea, sage, olive and forest green.

This rather famous actress and fitness guru also has a web site[7] entitled: *Goop.com.* Now it is interesting to wonder how anyone so petite and dainty can come up with such a title, but she did. And some of the Earth-friendly clothes on this site "range from a pricey $435 to an eye-bulging $1535."

However, one of the commentators on this news item seemed to come up with the right answer: Paltrow is somewhat green! Clearly not dark green, not a British Green, but a very very light green.

• Some Comments on how Green are Spain and Denmark.

Barack Obama has instructed us "'to think about what's happening in countries like Spain' as a model[8] for a U.S. future." But Spain is suffering "from enormous public debt incurred through programs like a mandated 'green economy'." Since then Obama has changed his instructions: "Think about what's happening in countries like Denmark." Of course Denmark is also not a useful energy model[8] as it's economy is also not robust, due in part to their huge investment in wind energy. The author cites a series of anecdotes, "referred to as 'the fairy tale of the windmills'", including:

· forcing people off their land;

· Clear cutting 15, but soon 30 square kilometers of forest to put up more windmills;

· locating some of these mills next to pig farms, that motivates these pigs to produce air pollution.

• **Anthony Blunt (1907 - 1983)**. Blunt[9.1] was a British subject with strong ties to British society, military, academy and the Royal Family. In 1952 he confessed to being a Soviet spy. For cooperation with the authorities, the government granted him full immunity and agreed to keep his spying career an official secret for many years. He was able to return to his former life, at least in part, and led a highly productive career as an art historian. His memoirs were finally released on July 23, 2009.

What follows are extracts from an essay[9.2] where the columnist uses and reacts to these memoirs. Blunt's memoirs "reveal a different age, one in which fascism and communism were locked in a seemingly definitive battle for souls." He saw a "'religious quality' of the enthusiasm for the left among the students of Cambridge." This columnist went on: "there is only one ideology in today's developed world that exercises a similar grip." If Blunt were young today he would not be red; he would be green."

This writer went on: "We are at the early stage of the green movement." Later he asks for the readers indulgence as he uses some historical determinism: "We, the peasants, are failing to rise up and embrace the need to change." - - - "Our intransigent refusal to choose green will be met by a new militancy from those who believe we must be saved from ourselves. "Ultra green states cannot arise without some form of forced switch to autocracy; the dictatorship of the environmentalists."

• **Some Comments on** *The Green Agenda*

Is the 'Green Agenda' appropriate in poor cities? This MIT paper[10] reports on conflicting agendas: helping the poor , the Brown Agenda versus protecting the future, the Green Agenda. A brief comparison of these two is noted in the table below.

Key Features	Brown Agenda	Green Agenda
First Order Impact	Human Health	Ecosystem health
Timing	Immediate	Delayed
Scale	Local	Regional and global
Worst affected	Lower income groups	Future generations
Typical proponent	Urbanist	Environmentalist

The Brown and the Green Agendas

The writer concludes that in theory both agendas could be addressed. However, "in practice, especially in poor cities, the more obvious priority is to assist in the development of locally-driven environmental initiatives."

Closing Thoughts

One can only second the council of Kermit the Frog. First, there are so many voices out there commanding one to "Go Green." The message is implied that all one has to do to be great, to be OK, is to "Go Green." And as a consumer, all one has to do is "buy green." I suspect I preferred the cheerleaders at various football games that commanded one to "Go Red, or Go Blue or Go Burnt Orange." So it would appear that this green command is not much more than a cheerleader's cheer.

Business owners face the same command. For example several times a week I pass a bank that named itself the Green Bank. This, of course, means this is a great bank, and an OK bank. A recent Fortune Small Business cover[11] declares "Green Power." It states that Earth-friendly isn't just a slogan — it's a profit center." Then the first page of the write-up states that with help from Washington, a new crop of entrepreneurs is building pragmatic, planet-friendly enterprises.

Universities are not going to miss this tidal wave. A tiny piece of evidence is a post card I received from Rice U. announcing that their Baker Institute for Public Policy was "going green!"

Ditto for the media. For example the CBC have come up with an initiative called a *Million Acts of Green*. In June 2010, they announced[12] they had counted 1,798,185 acts, and saved 106,114,018 kg of greenhouse gases. The CBC claimed this was a big hit. But they forgot to mention that the GHG savings amounts to a mere 0.0004 percent of total CO_2 emissions from the use of fossil fuels..

171

But what shade of green? Perhaps the shade of the green bud of the **Amorphophallus titanum** is getting close. This would be a very fine choice, particularly if that shade is reflective of the aroma of it's flower, the corpse flower – known as "the worlds stinkiest[13]." As its name warns, "the corpse flower is a smelly thing, with the withering stench of rotting flesh."

Surely if there can be over a million acts of green there must be a million shades of green. So which shade, out of these million or so shades, should be reserved for the Gorons, the ultra liberals, the far-lefters and the global warmers? My selection will be *"Amorphophallus titanum* green," (ATG). However, since ATG is such a difficult color to spell, and since it has so many syllables, an alias is needed. This must be a name that is a bit simpler, but captures the same aroma of ATG, and that name will be Gangrene Green (GG).

Hence the answer to the key question of this book—on "How Green are the Gorons?"—is like Gwyneth Paltrow: a very light green. It is not a very deep green, and it can only be one shade, out of a million or so shades. And that shade is ATG, alias GG. And this shade is not a very healthy green.

This book has been about energy issues and the conviction that we can, indeed that we must use fossil fuels, including coal. This book has been about environmental issues and the possibility of climate change. It has been about politics and the tidal wave of propaganda within which we live.

It has also been about the space alien community, primarily the Gorons, and their attempt to kidnap the color green. Now I may have been a bit harsh on this group. Indeed as noted in the Preface, I want the Goron name to be preserved. Hence, in an effort to leave this subject on a positive tone, I will close this book with an ode to these interstellar travelers.

An Ode to the Gorons and Morons of the Space Alien Community

> Oh what a great bird is the Goron.
> His brain holds less than a Moron.
> He can hold in his fist.
> An incredible list.
> Prepared by Prince Albert the Bore-on.

References and Notes

(1) Heinlein, Jr., Robert, A., *The Green Hills of Earth*, a short story, with the poem of the same title, Oct 20,1999. See: http://www.cs.rice.edu/~ssiyer/minstrels/poems/241.html.

(2) As reported on Johnny Cash - the Streets of Laredo lyrics.

(3) For the lyrics to: *It's Not Easy Being Green* see: http://kids.niehs.nih.gov/lyrics/green.htm.

(4.1) Simon, Stephanie, *Green vs Growth: The Battle Rages On*, Wall Street Journal, April 17-18, 2010. This article also includes several partial clippings from their archives, including: *Shades of Green Eight of 10 Americans are Environmentalists, At Least So They Say*, August 2, 1991.

(4.2) Huber, Peter, *Hard Green - Saving the Environment from the Environmentalists*, Basic Books, New York, NY, 1999.

(4.3) Ruckelshaus, William, *A New Shade of Green*, Wall Street Journal, April 17, 2010.

(5a) Mouawad, Jad, *ExxonMobil - Green is for Sissies*, New York Times, November 16, 2008.

(5b) Wadler, Joyce, *Green Guilt. Even the most committed say they commit environmental sins*, New York Times, September 30, 2010.

(6) Rubin, Paul H., Environmentalism as Religion, Wall Street Journal, April 22, 2010. The writer is a professor of economics at Emory University, and author of - *Darwinian Politics: The Evolutionary Origin of Freedom*.

(7) Shannon, Meg, *How Green is Gwyneth Paltrow*, as reported on Fox News, July 23, 2009.

(8) Horner, Christopher, *Obama's Model 'Green' Country? Denmark Evicts Citizens, Clear Cuts Forests for Windmill Space*, Pajamas Media, May 24, 2010.

(9.1) See: http://en.wikipedia.org/wiki/Anthony_Blunt

(9.2) Senior, Antonio, *Blunt warning about greens under the bed*, Times OnLine, July 24, 2009.

(10) *Is the 'Green Agenda' appropriate in poor cities?*, as posted at webmit.edu, December 5, 2000. Key table in this report was by McGranahan, G. and Satterthwaite, D., *Environmental Health or Ecological Sustainability: Reconciling the brown and green agendas in urban development*.

(11) *Green Power*, Fortune Small Business, April 2009.

(12) See: http://www.cbc.ca/green/.

(13) Huber, Kathy, *Your nose will know*, Houston Chronicle, July 8, 2010.

APPENDIX

Key Abbreviations, Acronyms and Chemicals

AAS — Arnold Alois Schwartzenegger

AASLs - Anthropogenic aerosol chemicals such as SO_2 and SO_x.

ACVR — American Center for Voting Rights.

AE — Alternative Energy.

AEI — American Enterprise Institute.

AGU — American Geophysical Union.

AGW — Anthropogenic Global Warming.

AIT — An Inconvenient Truth.

ANWR — Arctic National Wildlife Refuge.

AMS — American Meteorological Service.

AO — Atlantic Oscillation.

'AR' — 'Acid Rains.' Refers to the acid satire in my first book.

AP — Associated Press.

ASLS — Aerosols.

ATG — Amorphophallus titanum green

BC — British Columbia.

BCF — Billions of Cubic Feet

BHO — Barack Hussein Obama.

BPA — Bonnyville Power Administration.

BP — British Petroleum.

BPD — Barrels per Day.

CAFE — Corporate Average Fuel Economy.

C&T — Cap and Trade.

CASLs - Natural aerosol chemicals, such as SO_2 and SO_x.

CBC — Canadian Broadcasting System.

CBS — Columbia Broadcasting Service.
CCC — Catastrophic Climate Change.
CDC — Center for Disease Control.
CEI — Competitive Enterprise Institute.
CEO — Chief Executive Officer.
CERA — Cambridge Energy Research Associates.
CFCs — An abbreviation for chlorofluorocarbons, a complex mixture of chemicals containing
carbon, chlorine, fluorine and hydrogen.
CGCMs — Coupled General Circulation Models.
CH — Carbon Hydrogen free radical.
CH_4 — Chemical formula for Methane.
$(CH_4)_2$-S — Chemical formula for dimethyl sulfide.
CIA — Central Intelligence Agency.
CNN — Cable News Network
COP — Council of the Parties.
COS — Chemical formula for Carbonyl Sulfide.
CO_2 — Chemical formula for Carbon Dioxide.
CO_2e — Expression for Carbon Dioxide equivalents. A calculated value based on the levels of all
greenhouse gases in the atmosphere, and their relative warming characteristics.
CO — Chemical formula for Carbon Monoxide.
Cpt1, Cpt2, Cpt3, Cpt4 — Components, defined by individual equations. See Reference 11, §3.2.

DDT — Dichlorodiphenyltrichloroethane.
DNA — Distinguish natural from anthropogenic causes.
DNA — Deoxyribonucleic acid.
DASLs — Dusts aerosols.
DMYs — Dummy variables are often included in econometric models where one suspects a factor exists, but cannot define it. The presence or absence of volcanic activity, such as the Mt Pinnatubo eruption, might be an occasion to specify a dummy variable.
DW-H — Deepwater-Horizon.
1-D — One dimensional climate model.
3-D — Three dimensional climate model.
24D — A herbicide. Precise notation is 2,4-D for 2,4-dichlorophenoxyacetic acid.

EAIS — East Antarctic Ice Sheet.

ECD — Equivalent Carbon Dioxide. See also CO_2e.

EDF — Environmental Defense Fund.

EMA — Earth Motion Anomalies around the sun such as eccentricity, tilt and wobble.

EMSs — Earth Magnetosphere variations.

EN — El Niño.

EN/LN — El Niño/La Niña.

EPA — Environmental Protection Agency.

EROEI — Energy Return on Energy Invested.

FBI — Federal Bureau of Investigation.

FBK — Feedback variables are often included in econometric models, and could apply here too.

FCCC — Framework Convention on Climate Change.

FDA — Food and Drug Administration.

FDR — Franklin Delano Roosevelt.

FERC — Federal Energy Regulatory Commission.

G — Gore vote. See Unabomber quiz.

GACs — Global Air Circulations.

GBD — Ground based temperature data.

GH — Gerhard Herzberg.

GPM — Gallons per minute.

GCM — General Circulation Model.

GCMI — George C. Marshall Institute.

GCRs — Galactic Cosmic Rays.

GG — Gangrene Green.

GHE — Greenhouse Effect

GHGs — Greenhouse Gases. Includes CO_2, CH_4, CFCs, N_2O and stratospheric O_3.

GLA — Glacial Lake Agassiz

GPM — Gallons per minute.

GSCs — Global Sea Circulations.

GW — Global Warming.

GW — Gigawatt, a basic unit of electrical capacity, a billion watts.

gwh — gigawatt hour, a basic unit of electrical energy, a billion watt hours.

HA — Habibullo Abdussamatov.
HBO — Home Box Office.
HC — Hurricane.
HG — Head Goron.
Hg — Chemical symbol for Mercury.
H, H & H — Hurricanes Hardly ever Happen.
HT — Hendrick Tenneckes
HW — Holland & Webster.
H_2O — Chemical formula for water.
H_2O_L — Chemical expression for water in the liquid phase.
H_2O_V — Chemical expression for water in the vapor phase.

IPCC — Intergovernmental Panel on Climate Change. Reports jointly to UNEP and WMO.
IR — Infra Red Rays.
ISO — Independent System Operator.

J — Journal.
JFK — John F. Kennedy.
JFK* — John F. Kerry.
JWT — J. W. T. Spinks.

K — Kilo, a thousand.
KP — Kyoto Protocol.
KW — Kilowatt, a basic unit of electrical capacity, a thousand watts.
kwh — kilowatt hour, a basic unit of electrical energy, a thousand watt hours.
KY — Kiloyears.
KYBP — Kiloyears before present.

LA — Louisiana.
LASLs — Land based organic aerosols.
LIA — Little Ice Age.
LBNL — Lawrence Berkeley National Laboratory.
Li — Chemical symbol for Lithium.
LNG — Liquefied Natural Gas.

M — Mega, a million.

MASLs — Marine organic aerosols: Carbonyl Sulfide (COS); Dimethyl Sulfide CH_4-S-CH_4.

m — Meter.

MA — Massachusetts.

MBPD — Millions of barrels per day.

mbp — meters below present.

MDN — Midland Daily News.

Mg — Chemical symbol for Magnesium.

MIT — Massachusetts Institute of Technology.

MMO — Miriam M. Oliphant.

MRA — Multiple Regression Analysis.

MW — Megawatt, a basic unit of electrical capacity, a million watts.

mwh — megawatt hour, a basic unit of electrical energy, a million watt hours.

MWP — Mediaeval Warming Period.

Na — Chemical symbol for Sodium.

NACC — National Assessment of Future Impacts of Climate Change.

NAO — North Atlantic Oscillation.

NaS — Chemical formula for Sodium Sulphide.

NAS — National Academy of Sciences.

NASA — National Aeronautics and Space Administration.

NB — New Brunswick.

NCAR — National Center for Atmospheric Research.

NGO — Non Governmental Organization.

NIMBY — Not In My Back Yard.

N_2 — Chemical formula for Nitrogen.

N_2O — Chemical formula for Dinitrogen Ooxide.

NO_2 — Chemical formula for Nitrogen Dioxide.

NO_x — Chemical expression for a family of Nitrogen oxides.

NOAA — National Oceanic and Atmospheric Administration.

NP — Nancy Pelosi.

NRC — National Research Council.

NYMEX — New York Mercantile Exchange.

NZ — New Zealand.

OCS — Outer Continental Shelf.

OECD — Organization for Economic Cooperation & Development.

O&GJ — Oil & Gas Journal.

OGCMs — Ocean General Circulation Models.
OPEC — Organization of Petroleum Exporting Countries.
O_2 — Chemical formula for Oxygen.
O_3 — Chemical formula for Ozone.
OTC — Offshore Technology Conference.

P — P, as used in alliteration.
P, P & P — Pimps, Prostitutes and Propagandists.
PA — Pennsylvania.
PCA — Principle Component Analysis.
PG&E — Pacific Gas and Electric utility.
PNAS — Proceedings of the National Academy of Sciences.
PP — Peter Principle.
ppm — Parts per million.
PURPA — Public Utility Regulatory Policy Act.

RA — R. Anthes.
R & D — Research & Development.
RNLA — Republican National Lawyers Association.
RPJr — Roger Pielke Jr.
RC — Rachel Carson.

SCE — Southern California Edison.
SD — Sustainable Development.
SDB — Satellite Data Base.
SDG&E — San Diego Gas and Electric.
SEIU — Service Employees International Union.
SEPP — Science and Environmental Policy Project.
SI — Sherwood Idso.
SL — Sea Level.
SLR — Sea Level Rise.
SO_2 — Chemical formula for Sulfur Dioxide.
SOAs — Solar Output Anomalies would include Sunspots, the Sunspot Cycle Length, Solar magnetic activity, the Solar Wind and so forth.
SS — Slaughter Solution.
SS — Stephen Schneider.
SSAs — Solar System Anomalies: orbital tilt, asteroidal and interstellar dust; Galactic Influences. Cosmic rays appear to be an important factor.
SST — Sea Surface Temperature.

THC — Thermo-haline circulation.
TR — Theodore Roosevelt.
TRA — Theodore Roosevelt Association.
TS — Tropical Storm.
TWTW — The Week That Was, an Internet newsletter, by SEPP.
245T — A herbicide. Precise notation is 2,4, 5-T for 2,4,5-trichlorophenoxyacetic acid.

U — Unabomber vote. See Unabomber Quiz.
U — University.
UB — Unabomber.
UCG — Unconventional Gas Resource.
UCAR — University Corporation for Atmospheric Research.
UHI — Urban Heat Island.
UN or U. N. — United Nations.
UNEP — United Nations Environmental Program.
USFWS — United States Fish & Wildlife Service.
USSR — Union of Soviet Socialist Republics.
UV — UV Rays.

W — Watt, a basic unit of electrical capacity.
WAIS — West Antarctic Ice Sheet.
WMO — World Meteorological Organization.
W/m^{-2} — Watts per meter squared.
WR — William Ruckelshaus.
WSJ — Wall Street Journal.
WUWT — What's Up With That, an Internet newsletter by Anthony Watts.
WWI — World War I.
WWII — World War II.

x — Concentration of Carbon dioxide. See Reference 11, §3.2.

NOTES

NOTE 1 — A Tribute to Eric Stanley Westbrook

As noted at the start of this book that it was dedicated to my father: Eric Stanley Westbrook.

He was born on April 14th 1899, at Balgonie, Saskatchewan, and passed away on November 7th 1963, at Saskatoon, Saskatchewan

My love for nature stems from my father. First of all he was a master gardener. Gladiolas were his specialty, but he also took great care with his Dahlias, Peonies and tomato plants. While I have never been a gardener, I grew up in a house full of gladiolas and peonies, and a love for fresh tomatoes. And I have frequently expressed my awe at the grandeur of the botanical bounty of Earth. For example in Texas: the Blue Bonnets and the Indian Paint Brush wild flowers; the Azaleas and the Hibiscus and their panorama of color; the Bougainvillea and the Crepe Myrtles, and their incredible long season in bloom; and the Live Oaks and their magnificent profile and the Pin Oaks and their great leaf.

He frequently got his family out to the local parks: first to the parks along the South Saskatchewan River, inside the city; then the less established parks along the river, outside the city; and finally the Forestry Farm at the edge of the city. Although this park was indeed a forestry farm, it was nonetheless a lovely park, with magnificent lawns, gardens and trees. Today it now includes the city zoo. He would also take us out to the University farm, to see the baby lambs, and the huge bulls.

As all fathers he wanted me to participate in sports. His interests were baseball, curling, hockey and golf. Hockey and curling were a lost cause for me. Baseball and golf were a better fit. We played catch frequently even though my glove was pretty pathetic, and my hand ended up rather raw. But I eagerly awaited the next outing. We also went to many a ball game to

181

watch some pretty good semi-pros play. These included the best Canadians in the area such as Gordie Howe, the great hockey player, plus several *imports*, primarily from U.S. colleges. I can recall many double headers, where we would pack a lunch and go and cheer in the bright prairie sun. As for golf we didn't seem to have as much *chemistry*. I was the worst student ever, and I used to think my father was the worst teacher. But somehow I still picked up a lot that I utilize today. We walked around many a prairie golf course inevitably looking for my ball. Even today, as little golf as I am able to muster, I love playing the courses in Houston, never keeping score, but enjoying the grandness of Mother Nature, and the very rare good hole. My sports ultimately turned out to be basketball and tennis, sports that my father never played.

Next, he did his best to get his family to the superb lakes that started about 150 miles to the north; to the Manitoba Escarpment area, 350 miles to the east/southeast; and to the Rocky Mountains, 500 miles to the west. None of these trips were easy. First there were many miles of gravel roads to deal with, always awful, sometimes terrifying. Then there were economic realities to deal with until the depression had run its course. And, finally, there was gasoline rationing to deal with until World War II (WWII) had run its course.

I was probably 12 before my first trip into the north, and I simply couldn't believe the beauty of this lake country and the Boreal Forest. I was probably 14 before my first trip to the Rockies, and I simply couldn't believe the majesty of these mountains. However, this trip was not without its trauma as we had eight flats over a period of three days. These flats were the unintended consequence of having a picnic lunch, near the Alberta border, where we parked over a spot where the local rancher had dropped dozens of fencing nails. The last of these flats was at the top of the now abandoned "Big Bend Highway" deep into the mountains of British Columbia, near the start of the Columbia River, and about 50 miles from the nearest service station. We made it to that station and on to Vancouver, Seattle and back to Grand Coulee Dam. I didn't know it then, but this was the start of my interest in the Columbia and other great rivers.

It was not until my father passed away that I came to realize he was not what one would call well-off. He had provided good shelter for his family. He had provided for his wife and family in the event of his early death. And somehow he guided us through the dust bowl, the depression and WWII. The company he worked for survived over this time span and he kept his job all through this period of trauma, walking a tight rope over an abyss of economic disaster. I don't know what nightmares he must have had, but somehow he coped.

He never had any opportunity for a real education as WWI came along early in his life. He served in the 1st Canadian Division, 77th Overseas Field Battery, from 1916 to 1918. Further, his father was a man who was brought up in the very strict British code of discipline. He applied that to his family, along with a very strong work ethic. Additional education for my father, after his military service, was simply not possible for a variety of reasons. And any moral support or accolades from his father probably never occurred. Yet he survived.

The pathway my father followed was hard, and hard on him, and contributed to a variety of shortcomings. On looking back I can now see I tended to magnify his shortcomings and to minimize his many good points. I can now look back and see many times where I was not as generous to my father as I could have been and should have been, for which I now deeply regret. As such this book is dedicated to his memory.

NOTE 2 — A Tribute to Rutherford Aris

As noted in the Acknowledgements the two most important professors in my development both obtained their PhDs from the University of London. What follows is background material on their careers and their rather incredible accomplishments, followed by a tribute to R. Aris.

• Rutherford Aris (Gus).

I attended Minnesota from September 1958 to June 1960 and obtained an M.S. in Chemical Engineering and an M.A. in Energy Economics, with Aris as my advisor on each thesis. I look back, with a great deal of fondness, to studying under his leadership.

The two thesis, that I was involved in at Minnesota, were on
· chemical reactor safety and stability analysis, and
· energy conservation in the process industries.

• J.W.T.Spinks (JWT).

JWT was head of the Department of Chemistry and Chemical Engineering, and Dean of Graduate Studies at the University of Saskatchewan, when I started there in 1951. In spite of being the head of the department, Spinks had the wisdom to give the compulsory chemistry class for all first year engineering students. In short he was the number one salesman for his department. He was an outstanding teacher and scientist in his own right. However his salesmanship capability was one that he utilized most effectively throughout his career. He later rose to become president of this university, a position he held for 16 years.

JWT became an assistant professor of chemistry at the University of Saskatchewan in 1930. He was all of 22 years old. The next three years were very busy ones for him: organizing lectures, starting a research program, meeting many new people, and getting to know his new city, province and country. However, by 1933, the economic conditions became so bad that all the unmarried professors were told to take a year off.

JWT headed to the Technische Hochschule at Darmstadt, Germany in September of 1933, to study under a Gerhard Herzberg, (GH) about four years his senior. GH was a German, a physicist, an astronomer and a specialist in the field of molecular and atomic spectroscopy. He was one of the founders of this area of study and ultimately wrote six books on this subject that are considered the *classics* in this field.

Now, while conditions were bad in Saskatchewan, they were becoming terrible in Germany, with Hitler receiving 90 percent of the vote in November of 1933. Yet in spite of the turmoil, the *chemistry* between Spinks and Herzberg was outstanding, and the research went exceptionally well. Spinks and Herzberg were able to get eight publications out of a years research.

• To North America

Spinks returned to Saskatchewan in September of 1934. Shortly after that Herzberg contacted him to see if he could get the president of the university to write a recommendation to the University of Toronto. Thousands of German scientists were looking to find new employment, which was now an extremely difficult, indeed, almost an impossible endeavor. Even though JWT was no more than a junior member of the chemistry department, and just barely back from a forced leave of absence, he convinced the president to act. Letters were written in January 1935 to both the University of Toronto and also the National Research Council (NRC) in Ottawa. The end of each letter noted, if circumstances did not exist for a suitable position, that somehow the U. of Saskatchewan—whose key financial assets at that time were not much more than a file cabinet full of IOUs— would make him an offer. That is what happened and the Herzbergs escaped from Nazi Germany, and came to Saskatchewan in the fall of 1935, with all of $2.50 left in their possession.

Herzberg served the University of Saskatchewan for ten years as a professor of physics, then moved to the University of Chicago and it's renowned Yates Observatory for three years. He then returned to Canada, at the NRC in Ottawa, to establish a major new laboratory in spectroscopy.

There is little question that without the recognition by JWT of the genius represented by Herzberg, plus his enthusiasm for, and his bond of friendship with this man—and his sales efforts to the key leaders at the University of Saskatchewan, the provincial government, and elsewhere— Mrs. Herzberg, and probably Dr. Herzberg, would have been sent to *Auschwitz*.

Herzberg won the Nobel Prize in chemistry, in 1971, for his contributions to the knowledge of electronic structure and the geometry of molecules, particularly what are known as free radicals.

JWT's specialties were physical chemistry, photochemistry, spectroscopy and radiation chemistry. He was one of the founders of the field of radiation chemistry. He pioneered the use of radioactive tracers and made significant contributions to agriculture, medicine and chemical research. In 1982 he was inducted into The Saskatchewan Agricultural Hall of Fame in part for his work on nutrient uptake and the improvement of the effectiveness of fertilizers.

During WWII he was assigned to the Canadian Operations Research (OR) effort, an activity that used logic, mathematics and statistics on operational problems. He was awarded the Member of the Order of the British Empire for his work on search and rescue procedures for missing aircraft.

Later JWT became part of the Canadian atomic energy project. His interest in, and confidence on nuclear power contributed towards my interest in this energy alternative. Indeed, I came within a whisker of accepting an employment offer at Canada's nuclear research center.

However his activities in OR also created a major interest for me in that field. This was the area in which I focused on at graduate school, and ultimately spent several years of my career in direct practice, and utilized the approach and discipline in all later areas of activity.

• A Tribute to Rutherford Aris.

I started to think of an analogy here early on, what with the two most important professors to me, coming from Great Britain in their early twenties. Both obtained their PhD from the University of London. Both were heavily involved in physical chemistry and chemical engineering. Both were fluent in languages: Aris published a technical paper once, in Latin; JWT translated the first volume of Herzberg's *magnum opus* – Molecular Spectra and Molecular Structure. And both were religious.

More recently I have started to see a second and more important analogy. This started with the observation that JWT studied complex chemistry under Herzberg, four years his senior. Four years was also the age difference between myself and Aris. Herzberg concentrated on complex chemical and physical research, which led to the Nobel Prize. Aris concentrated on complex chemical engineering research which led to many honors. While these honors did not include the Nobel Prize, in my mind, Aris is my *Nobel Prize* winning scientist, engineer and professor.

From Chapter 2.2

NOTE 3 — The Econometric Model Approach to Temperature Models.

This approach—namely the econometric model specification approach—highlights the challenge involved in the analysis of the Earth's temperature and in modeling the climate. Surely, no simple statistical model—specifically a linear regression model, sometimes called a least squares model—could contain all these variables.

Indeed the complexity involved rapidly leads to the abandonment of statistical models and the movement to the use of huge computer, simulation models. These are known as the so-called general circulation models (GCMs). While there has been some success with the GCMs, there also have been significant failures. The GCMs will be covered in Chapter 3.3.

In summary the subject of temperature analysis is a most complex story. In addition to the material covered in Chapter 3.3, comments will be made on this subject throughout this book and on the so-called hockey stick profile, another high profile and highly controversial temperature model.

References and Notes for NOTE 3

Such *model(s)* are expressed via the following convention:

T = a Function of {x, y, z, etc}.or

T = F(x, y, z, etc).

where T would be temperature and x, y, z, etc are variables to be defined.

The approach used by the HG is based on a very simple (and unstated model) namely:

$$T = F\{[GHGs]\},$$

where [GHGs] represents a set of greenhouse gas variables. For the HG these would be the concentration of CO_2, and possibly methane, (CH_4), in the atmosphere.

However, when all the variables that may be involved are added, the final model specification becomes incredibly complex. It might look like something as follows.

$$T = F\{- - - - - [GHGs] - - - - -\},$$

where the blanks represent 14 -15 other sets of variables, perhaps a total of 100 or so variables as noted below.

Econometric Model Specification.
T = F{ [AASLs], [CASLs], [DASLs], [DMYs], [EMAs], [FBKs], [GACs], [GHGs], [GSCs], [LAGs], [LASLs], [MASLs], [SOAs], [SSAs], [Other] }.

Major Categories	Examples of Groupings of Potential Climate Variables
Aerosols	AASLs - Anthropogenic aerosol chemicals such as SO_2 and SO_x
	CASLs - Natural aerosol chemicals, such as SO_2 and SO_x
	DASLs – Dusts, including pollens
	LASLs - Land based organics such as terpenes and isoprenes
	MASLs - Marine based organics such as Carbonyl Sulfide (COS) and Dimethyl Sulfide $((CH_4)\text{-}S\text{-}(CH_4))$
	Water droplets - H_2O_L
Greenhouse gases	Water vapor - H_2O_V, the most important GHG
	Basic gases - CH_4, CO, CO_2, O_3, N_2O, NO_2 and NO_x
	CFCs - the chlorflourocarbons. (See Appendix).
Geophysical items	EMAs - Earth Motion Anomalies around the sun - eccentricity, tilt and wobble
	EMSs - Earth Magnetosphere variations
Astrophysical items	SOAs - Solar Output Anomalies would include Sunspots, the Sunspot Cycle Length, Solar magnetic activity, the Solar Wind and so forth. It is interesting to note that while the warmers make much to do about the increase in atmospheric concentration of CO_2 by 30%, they have nothing to say about the sun's magnetic flux, that has doubled during the last 100 years[2].
	SSAs - Solar System Anomalies: orbital tilt, asteroidal & interstellar dust.
	Galactic Influences - Cosmic rays appear to be an important factor. However, variation, in what seems to be a constant flux, appears to be through modulation by the solar magnetosphere and by the EMSs
Circulations	GACs - Global Air Circulations
	GSCs - Global Sea Circulations. These two fields would include such phenomena as the El Nino, the Southern Oscillation; and surface ocean, and deep thermo-haline circulations.
Other variables	Location - Hemisphere, Latitude, Altitude
	Timing - Summer/Winter, Night/Day.
Econometric Techniques	DMYs - Dummy variables are often included in econometric models where one suspects a factor exists, but cannot define it. A similar situation would exist here. The presence or absence of volcanic activity, such as the Mt Pinnatubo eruption, might be an occasion to specify a dummy variable, depending on the subject under study and the time interval involved.
	FBKs - Feedbacks are often included in econometric models. A similar situation would exist here.
	LAGS - Many variables are included on a lagged basis in econometric models. A similar situation would exist here.

A Summary of the Variables with Potential Influence on Our Climate.

From Chapter 3.2

NOTE 4 — On the debate between Newell, Doplick, Idso and Schneider.

<u>Prologue.</u> By 1979 most experts remained confident that the radiation models used as the basis for GCMs were fundamentally sound, so long as they didn't "push the models too far." However, some scientists raised concerns. Reginald Newell and Thomas Dopplick questioned [1] the basic 1-D models. They pointed to a weakness in the GCM predictions that

an increase in CO_2 levels would bring a large warming. They noted that such a prediction depended crucially on the assumptions about the way a warming atmosphere would contain more H_2O_v. They suggested that the popular climate models might over estimate the temperature rise by an order of magnitude. They cast major doubt on whether scientists really understood the GHE at all.

In 1980 Sherwood Idso joined this dissent. Idso was initially viewed by Weart as a respected scientist at the U.S. Water Conservation Laboratory in Arizona. His work was felt to offer particularly interesting arguments as he sought[3, 4] out "natural experiments" where climate sensitivity could be measured. Idso published arguments that GHG gas emissions would not significantly warm the Earth. Better still, by *fertilizing* crops and trees, the increase of CO2 would bring tremendous benefits.

The estimates by Idso. He joined the attack on the models. He argued he could determine climate sensitivity to additional GHGs by applying elementary radiation equations to several basic "natural experiments." He pointed out, for example, that one could look at the difference in temperature between an airless Earth and a planet with an atmosphere, or the difference between Arctic and tropical regions. He also compared, as did others[5], the conditions on Venus and Mars versus Earth. Since the temperature differences, in all the above cases, were in the tens of degrees area, he concluded that the smaller perturbation, that came from doubling CO_2, must cause only a negligible change, a tenth of a degree or so, perhaps as much as 0.4ºC. Clearly Idso's estimates were an order of magnitude below the conventional wisdom of mainstream scientists. He published these ideas and created a vehement controversy. Starting in the late 1970s and extending to the late 90s, Idso studied eight natural events. Seven of these are listed below.

Case I - Phoenix area. Over ~ 45 days a summer monsoon occurs, resulting in fluctuations in the atmospheric vapor pressure. Significant changes in thermal and solar radiation fluxes result.
Sensitivity estimate 0.173 ºC/Wm^{-2}

Case II - Phoenix area. A significant redistribution in the vertical dust concentration occurs between summer and winter. In winter the dust is held to much lower levels. This phenomenon does not alter the transmittance of the atmosphere for solar radiation, but it increases down-welling flux of thermal radiation.

Sensitivity estimate 0.173 °C/Wm⁻²

Case III - 81 U. S. Sites. The annual cycle of surface air temperatures is caused by the annual cycle of solar radiation absorption at the Earth's surface. A total of 81 sites could be divided into interior and Pacific rim locations. Used interior value for land area, and rim value for ocean areas, and estimated global weighted average.

· Land sites sensitivity estimate 0.171 °C/Wm⁻²
· Rim sites sensitivity estimate 0.089 °C/Wm⁻²

Global average sensitivity estimate 0.089 °C/Wm⁻²

Case IV - Global. Compared temperature for Earth with an atmosphere, to Earth without. Sees a ~33.6 °C warming. This is sustained by a thermal radiative flux of ~ 348 w/m². Flux would be zero without an atmosphere. **Quotient is 0.097 °C/(w m⁻²).**

Case V - Global equator to poles (EtoP) temperatures. The EtoP temperatures are set by the EtoP totally surface absorbed radiant energy. Literature data can be used for each 5° latitude increment:

(1) average surface air temperatures;
(2) H_2O vapor pressures;
(3) latitude distribution of cloud cover and
(4) average surface solar absorbed radiation.

From (1) and (2) can get the clear sky atmospheric thermal radiation at the Earth's surface at the latitude midpoints.

From (3) could modify the clear sky flux of thermal radiation. Could now add both fluxes to get 18 annually averaged total surface-absorbed radiant energy fluxes.

Item (1) could then be plotted against these. Two key trends were seen. A global value could then be obtained by weighting these results by surface area (12 & 88%).

Trend 1 from 90 °NS to 63 °NS, slope - 0.196 °C/(w m⁻²); and
Trend 2 from 63 °NS to the equator, slope - 0.090 °C/(w m⁻²).

Global value of 0.103 °C/(w m⁻²)

Case VI - Venus/Mars. A comparison of the GW situation on these two planets is of interest.

Planet	Global Warming °C	CO2 Level %	Pressure Bars	Partial Pressure Bars
Venus	500	96	93	400
Mars	5	100	0.01	0.01

Comparison of GW on Venus and Mars

A log-log plot of GW vs ∂P can be made. This permits estimate of GW on Earth at 300 and 600 ppm CO_2.
Gives a delta of 0.4 °C

Case VII - Faint Sun. Obtained history for CO_2 for 3.5 B years. Assumed ~25% cut in solar luminosity over this interval. Calculated the GW needed to compensate for this lower luminosity. Plotted GW vs CO_2
Led to 0.4°C.

Epilogue. Estimates for climate sensitivity from these natural phenomena ranged from 0.09 to 0.40 °C/Wm^{-2}. These are an order of magnitude lower than values obtained from modeling. Once published, a major fight between Idso and the modelers resulted.

Stephen Schneider (SS) and other modelers rejected the work of Idso, Newell and Dopplick.They claimed that Idso, Newell, and Dopplick were misusing the equations. SS claimed that "Along the way Idso attacked the 'scientific establishment' for rejecting his theories." SS described Idso "as refusing to admit his error." Schneider and other modelers were "dismayed"[6] by Idso's claims. They described [7] Idso's analysis as incomplete, simplified and, worst of all, "failed to raise the state of the art of climate modeling." Idso replied that he was not a modeler and made no pretension of "raising the state of the art of climate modeling." Rather he noted that he much preferred the approach of one who deals in experiments.

While Idso cited eight "natural experiments"—three in his 1980 paper and the others a bit later—that surely cast some doubt on the level of warming that could occur. Schneider—Weart noted—claimed Idso's conclusions were "based upon various violations of the first law of thermodynamics." One should have jumped at this charge[8] that a PhD holder in physics would violate this law. This would be a charge that would surely need some justification, else it would be rather incredulous. However, Weart did not challenge Schneider. A long technical controversy between Idso and modelers followed for decades..

Another observation was on Idso's reaction to the point that GW will lead to more H_2O_V which will lead to more warming etc, etc, etc. Idso argued if one started at 15 °C, and had an initial GW of 0.25 °C, this would increase vapor pressure by 0.2 millibars, which in turn would add a further warming of 0.07 °C, and that warming would add a little more moisture, which would add a further warming of 0.01 °C. In total this would end up with a warming less than 0.3 °C.

References and Notes for NOTE 4

(3) Idso, Sherwood B., The Climatological Significance of a Doubling of Earth's Atmospheric Carbon Dioxide Concentration, Science 207: 1462-63, 1980.

(4) Idso, Sherwood B., CO2 induced global warming: a sceptic's view of potential climate change, Climate Research, 10: 69-82, 1998.

(5) Bryant, Edward, *Climate Process & Change*, Cambridge University Press, 1997.

(6) Schneider, Stephen H., et al, Letters: Carbon Dioxide and Climate, Science, 210, 6-7, 1980.

(7) Reference and Note (7) is reported above at the start of this NOTE 4.

(8) While working for a major petrochemical company, our European office was approached by two *con artests* claiming they could desalt sea water for pennies. Their data was sent to our head office, then turned over to me. Very few inputs were given by these *con artests*, but just barely enough to allow me to calculate their energy use. This was lower than the minimum energy required, as defined by the Second Law of Thermodynamics. Hence these cons were trying to sell us a process that would break this law. Violating thermodynamic laws is a tactic that one might expect from *con artests*, but not from scientists with PhDs.

NOTE 5 — A Debate between a British Scientist and three U.S. Physicists.

A recent news story in the Houston Chronicle reported on an interview[9] with a Dr. C. Rapley, of the British Antarctic Service. He asked two rhetorical questions. "Do you agree that physics is physics? And "If carbon is increasing how can you really deny there's going to be warming?" Rapley challenged the readers, in the arrogant manner that many Britishers are plagued with: if you really knew how physics works, you would *stop arguing on* GW.

As a means to approach this challenge, a rather simplified analysis of the 1990 IPCC assessment will be noted. This is from the book[10]: *Scientific Perspectives on the Greenhouse Problem*. The authors are:

· Robert Jastrow, Columbia PhD, ultimately formed the Goddard Institute for Space Studies (GISS);

· William Nierenberg (1919 - 2000), a former director of the Scripps Institute of Oceanography;

· Frederick Seitz, (1914 - 2008) Princeton PhD in solid state physics and later president of Rockefeller U.

These authors, each with a PhD in physics, viewed the GW, of 1.5 to 4.5°C for the next century—as cited by the 1990 report, as alarmist. They based their analysis only on <u>observational data</u>, and no computer modeling. Their analysis included:

(i) <u>assuming</u> the temperature increase—from pre-industrial levels ~1880 - 1990—was 0.3 to 0.6°C;

(ii) <u>assuming</u> this rise was all due to a 50% increase in GHGs from pre-industrial levels;

(iii) <u>assuming</u> a 100% increase in GHGs, from 1990 to 2100;

(iv) as openers, one could then see twice the warming, or 0.6 to 1.2°C (That would seem pretty logical – double the increase in GHGs gives double the warming. However, there is a well known logarithmic relationship[11] on warming—successive additions of GHGs will have a lower impact versus the preceding addition. However, these three scientists did not incorporate that into their analysis.);

(v) <u>assuming</u> a correction of 0.2°C for ocean thermal lag, would give a revised range of 0.8 to 1.4°C;

(vi) <u>finally assuming</u> an allowance of ± 0.4°C for natural climate variability, would give 0.4 to 1.8°C.

Their simple analysis was the basis for their conclusion that the IPCC was far too pessimistic, and represented a major exaggeration of the actual physical situation.

References and Notes for NOTE 5

(9) *Scientist rebuts global warming critics*, in an interview of C. Rapley, British Antarctic Service, Houston Chronicle, February 7,.2006.

(10) Jastrow, R., Nierenberg, W. and Seitz, F., *Scientific Perspectives on the Greenhouse Problem*, The Marshall Press, Jameson Books Inc., Ottawa, IL, 1990.

NOTE 6 — Hansen's Alarming Testimony of 2007 and the Idso's family response

Prologue. In 2007: James Hansen presented alarmist testimony[12] to the House of Representatives in April. Hansen's testimony was entitled *Dangerous Human-Made Interference with Climate* and was based on a paper of the same title[13].

The summary of his testimony includes the claim that our "climate is remarkably sensitive to global forcings – – –. "Huge natural climate changes, from glacial to interglacial states, have been driven by very weak, very slow forcings – – –." He goes on to further claim that today we are "applying much stronger, much faster forcing" as we add CO_2 to the atmosphere. Sherwood Idso and son Craig did not wait very long to provide a critique[14] of this testimony.

The Critique. Sherwood Idso and son Craig did not wait very long to provide a critique[14] of this testimony. In a 25 page critique on their web site—43 pages as a quality paper copy—they identified seven key subjects they took major exception to, and another 15 that deserved comment. They noted that if " there is any human enterprise that should be free of appeal to authority, it is science, where observation and impartial analysis are supposed to reign supreme." Their comments included:

(1) when the outcome of an ongoing scientific investigation is perceived to be a powerful catalyst for governmental action by the world's community of nations; and when

(2) the leading policy prescription for those actions is something akin to a massive restructuring of the way the energy that runs the modern world is produced, distributed and used; and

(3) especially if the policy is developed before all pertinent data have been acquired and properly analyzed; this principle can easily be forgotten. In such circumstances, and even more so if the subject being studied is extremely complex—such as how human activity will impact global climate centuries into the future— and when a divergence of views develops because of ambiguities in the observations and different methods of analysis, it is important that personal opinion be clearly differentiated from demonstrable fact. The Idsos argue that Hansen has failed broadly and deeply in that context.

Epilogue. What followed then were comments by the Idso's on Hansen's testimony:

(i) ice sheet disintegration, (ii) sea level trends, (iii) atmospheric methane concentrations,
(iv) climates of the past, (v) predicted warming-induced extinctions of terrestrial plants and animals,
(vi) the CO_2 induced preservation of terrestrial species, and (vii) CO_2-induced extinctions of coral.

They also flagged 15 additional areas: (i) positive vs. negative climate feedbacks, (ii) effects of drought on agriculture in a CO_2 rich world, (iii) sea level rise over the next hundred years, (iv) the adaptability of organisms to rising sea levels, (v) the "dangerous" level of atmospheric CO_2, (vi) the size of climate forcing due to a 2X CO_2, (vii) empirical evaluations of earth's climate sensitivity, (viii-x) the ability of man to control global climate, (xi-xiv) need to act now to cut CO_2 emissions, and (xv) role of morality over what to do re CO_2 emissions.

No attempt is made here to cover all of these 22 areas. A few comments are in order. For example they noted that when Hansen's testimony is compared with what has been revealed by the scientific investigations of a "diverse assemblage of highly competent researchers in a wide variety of academic disciplines, we find that he paints a very different picture of the role of anthropogenic CO2 emissions in shaping the future fortunes of man and nature alike than what is suggested by that larger body of work."

References and Notes for NOTE 6
(11) Hansen, James et al, *Global Climate Changes as Forecast by Goddard Institute for Space Studies Three Dimensional Model*, See Appendix B: *Radiative Forcings*, Journal of Geophysic Research, August, 1988. Hansen et al describe equations for nine forcings for the various GHGs. The equation for CO_2 is logarithmic. Let:

x = CO_2 concentration in the atmosphere, in PPMv.

Sum = (1 + Cpt2 + Cpt3 + Cpt4), where:
Cpt2 = 1.2* x;
Cpt3 = 0.005*x^2; and
Cpt4 = 1.0*10^{-6}*x^3.

ΔT = Ln(Sum).

Delta1 = $\Delta T(x)$ - $\Delta T(x-1)$ and
Delta2 = $\Delta T(640)$ - $\Delta T(315)$.

Radiative forcing, for changes in the atmospheric content of CO_2, can be estimated by the above formulas. The accuracy is about 1%, within the range of 315 to 1000 PPMv. Example follows.

X PPMv	Cpt2	Cpt3	CPT4	Sum	AT = Ln(Sum)	Delta1 °C	Delta2
315	378	496.1	31.3	906.4	6.81	-	-
340	408	578.0	39.3	1026.3	6.93	0.12	-
365	438	666.1	48.6	1153.8	7.05	0.12	-
590	708	1740.5	205.4	2654.9	7.88	0.08	-
615	738	1891.1	232.6	2862.7	7.96	0.08	-
640	768	2048.0	262.1	3079.1	8.03	0.07	1.22

A Numerical Example of Radiative Forcing Calculation

(12) Hansen, James A., *Dangerous Human-Made Interference with Climate*, Testimony before the U. S. House of Representatives, April 26, 2007.
(13) Hansen, James A., et al, *Dangerous Human-Made Interference with Climate*, (a) published in Atmos. Chem. Phys. Discussions., December 5, 2006, (b) revised March 29, 2007, and (c) published May 7, 2007.
(14) Idso, Sherwood and Idso, Craig, *Carbon Dioxide and Global Change: Separating Scientific Fact from Personal Opinion*, June 6, 2007. See: www. co2science.org.

From Chapter 3.3

NOTE 7 — Key Skeptic Inputs: Richard Lindzen

· 1989. Mallove provides a report of a speech[12] by Lindzen, entitled: *Has GW already happened? Problems with models, and the Complexity of the Problem*. He reports:
(a) the GHE does not seem to be as significant as suggested;
(b) there is time and need for research; and
(c) GW is a subject area where the uncertainties are "vast." Lindzen noted that "the most wonderful constituent is water with its remarkable thermodynamic properties. It's the obvious candidate for the thermostat of our system, and yet in most of these models, all water related feedbacks are positive.
· 1990A. In this paper[13], *Some Coolness Concerning Global Warming*, Lindzen notes that the idea that there may be problems in existing models

is not new to the meteorological field. Models usually have had difficulty in replicating key climate features—such as: mean global temperature, equator to pole temperature gradients, and the intensity and position of the jet stream—without what is called euphemistically "tuning." Here processes "not resolved" in the model are parameterized to bring the model into line with the observations.

The key area for major problems will most likely be with the interaction of the climate with water in all its phases. The remarkable thermodynamic properties of water "certainly lead to its acting as natures thermostat." Yet in major models all feedbacks between warming and water are positive, that is they add to the warming. In the absence of these positive feedbacks these models would give a warming only one fifth to one half of those claimed. This paper drew much attention, particularly a letter by an A. K. Betts entitled *GH warming and the tropical water budget*. This in turn received a reply by Lindzen

(a) Betts letter argues that the failure to stress the role of H_2O may be due to its short retention time compared to other GHGs.

(b) Lindzens reaction[14]: "I find this remark incomprehensible." The retention time of H_2O has nothing to do with its IR properties. "Indeed it is its short response time that peculiarly enables it to play a possible regulatory role."

· 1990B. In a paper[16] titled, *Some Uncertainties with respect to water vapors role in climate sensitivity*, Lindzen argues that it is futile to talk about climate change without a deep understanding of the behavior of H_2O_V, and our present knowledge of the behavior of H_2O_V is inadequate to this task. H_2O_V has the dominant role in the radiative budget of the troposphere through its impacts on short and long wave radiation and its ability to form stratiform clouds. Clouds are not only important in the IR, but are also the key determinant of the Earth's albedo. He addressed two areas of uncertainty in this paper:

(a) heat transport to higher latitudes and altitudes; and

(b) the response of H_2O_V in the upper troposphere to climate forcing. This property is now unmeasured and the parameterizations used in large models, are clearly wrong on physical grounds.

· 1993. In a National Geographic paper[17], Lindzen notes that model predictions of a large GW depend on large increases in CO_2, and mechanisms within the models that greatly amplify the climatic response to increasing CO_2. These mechanisms, namely positive feedbacks, depend on what is likely a severe mis-statement of the key physical processes: moisturization of the atmosphere and cloud formation. Indeed these processes may be acting in a manner opposite to what current models produce. Lindzen notes, that while the possibility that a large GW has not been dis-proven, it is without a meaningful scientific basis.

References and Notes for NOTE 7

(12) Mallove, Eugene F., *Lindzen Critical of Global Warming Prediction*, MI T Tech Talk, Sept. 27, 1989.
(13) Lindzen, Richard S. (1990). *Some Coolness Concerning Global Warming*,. BAMS, 71: 288-99.
(14) Lindzen, Richard S. (1990), *Response*, BAMS, 71: 1465-1467.
(15) Not assigned
(16) Lindzen, Richard S. (1990), *Some uncertainties with respect to water vapors role in climate sensitivity*, Proceedings of NASA Workshop on the Role of Water Vapor in Climate Processes, October - November, 1990.
(17) Global Warming Debate, Research & Exploration, A Scholarly Publication of the National Geographic Society, Spring 1993.

From Chapter 4.2

NOTE 8 — Extreme weather events

Erik Larson noted five of these events in his book Issac's Storm. Two of the five events listed below are HCs not too far from Galveston. Hence they are of interest as background inputs on Issac's HC education. However all five can be viewed as extreme weather events and are also of interest as inputs on such phenomena. Frequently *warmers* cite extreme weather events as evidence of GW. Yet such events have been with us for thousands of years.

(1) Hurricane, September 16, 1875 - "The storm raised an immense dome of water and shoved it through Indianola, pushing the waters of the Gulf and Matagorda Bay inland 'until for 20 miles the back country prairie was an open sea'". Storm took 176 lives.

(2) Hail and flash flood, August 1885 - a severe downpour near San Angelo, including hailstones "the size of ostrich eggs", killed hundreds of cattle and created a flash flood with: "An escarpment of water that Issac estimated to be 15 or 20 feet high".

(3) Hurricane, August 20, 1886 - This storm completed the destruction of Indianola. So many residents were killed that the survivors abandoned the town completely.

(4) Blizzard, November 1888 - this surprise blizzard destroyed 150 vessels off New England, and caused the death of 450.

(5) Blizzard, January 1899 - "this blizzard swept much of the South. Icebergs 10 feet high flowed down the Mississippi past New Orleans". This storm even hit Galveston and piled snow on it's beaches and drove water out of the Bay into the Gulf.

References and Notes for NOTE 8

(6) Larson, Erik, *Isaac's Storm*, , Crown Publishers, New York, 1999.

(6.1) Larson cited several other extreme weather events in his book. These are cited above.

NOTE 9 — Debates over the potential couple between GW and HCs.

What follows are key selections from several debates on these questions.

• **Input by** K. Emanuel, R. Anthes and R. Pielske Jr.

· July 31, 2005. Kerry Emanuel, at MIT, argued [29] that both the duration and maximum speeds have increased by 50%. He further argued that this is mirrored by the increase in SSTs.

· May 2006. In dual articles, both comments[30] by R. Anthes et al (RA); and a reply[31] by R. Pielske Jr. (RPJr) et al to these comments are reported.

RPJr methodically refutes all the criticisms cited by RA and concludes with a consensus statement by the WMO: "The research issues studied here are in a fluid state and are the subject of much current investigation. Given time the problem of causes and attribution of events of 2004 - 2005 will be discussed and argued in the refereed scientific literature. Prior to this happening, it is not possible to make any authoritative comment." RPJr disagreed as his critiques were rather strong.

• **Input by Holland & Webster**

Two references are noted.

· July 30, 2007. There are several references to a major paper[32] by Holland and Webster (HW), that argue there is a major increase in the frequency of HCs. However, one of these references[33] reports that NOAA itself has debunked the conclusions of this paper.

· August 7, 2007. RPJr noted there have been some strong words between an author on RPJr, a Chris Landsea[34] and HW. Landsea has called their research "sloppy." The key conclusions of this critique included:

(i) Results of HW depend crucially on a yet to be completed re-analysis. Hence bold public claims to absolute certainty are way overstated.

(ii) Extension of trends—from a still uncertain analysis—to claims of conclusive attribution to GHG emissions may be supported by other work, but it is nowhere to be found in HW.

(iii) The American public should not have increased concerns about increasing HC landfalls, due to the work in HW, as this paper does not even discuss landfalls.

• Sea Surface Temperatures.

This area has been targeted extensively as the culprit, all due to GW. I would argue this is another case where the Gorons are using very simple science to explain a very complex phenomenon. One common denominator of all the above studies by the *warmers* is the conviction that GW has occurred and the oceans have warmed. However, as noted earlier, Dr. O'Brien has reported that the oceans have not warmed in the HC formation region of the Atlantic.

Other comments on SST may be useful.

· September 12. 2005. In an interview[24] James Glassman asked O'Brien - was the reason the Gulf was so warm due to GW? O'Brien laughed. I laugh because the entire Gulf in August is over 90 degrees. The problem here was simply that Katrina spent too much time over warm water.

· May, 2006. Michaels[35] et al commented on SST trends, that between:

(i) 21.5 to 28.25 °C, for each 1 °C rise in SST, the maximum wind speed rose by 2.8 m/s (6.3 mph).

(ii) As SSTs rise above 28.25 °C the first category 3 or greater storms begin to appear. However, there is no significant relationship between SST and maximum winds, at higher SSTs.

Consider the path Katrina followed. As it approached the Bahamas it was a TS, heading N by NW. It then veered towards Florida heading mostly west. It hit Florida as a level 1 HC. It then veered W by SW, back over water, and right over the extra warm Loop-Current[37], in the Gulf of Mexico. It then rapidly grew to a level 5 storm. Here the Yucatan Current moves north between Cuba and Mexico, forms an upside down horseshoe before moving out of the Gulf via the Florida Current, between Cuba and Florida. Sometimes this horseshoe is fat, sometimes thin. Sometimes this horseshoe stays south near the Florida channel. Sometimes this horseshoe can meander northward, almost to the the northern Gulf Coast Finally this horseshoe can give birth to eddies that drift west. When a HC crosses the Loop Current it may intensify rapidly, as it feeds on its warmth. An October

2000 report[38] noted that three of the last five HCs, Gordon, Keith and Helene, crossed the Loop-Current. And in 1999, HC Bret, jumped from a level 2 to level 4, in just a few hours, as it crossed an eddy, off the coast of Texas.

• Other Variables.

There are many variables here, perhaps as many as listed in §2.3 for global warming itself.

· Steering Currents. We have mentioned steering currents several times in §4.2. These forces are primarily atmospheric and are crucial factors in forecasting any storm. Yet they are no easier to forecast than the loop-currents we noted above.

· Ocean Currents. In addition to the Loop-currents there are dozens of ocean currents that can play a factor on the global perspective.

· El Niño/La Niña.

· Surface Winds.

· Upper level winds.

· Wind Shear. The above three items all may contribute to the wind shear, that can tear a developing storm apart. The subject of vertical wind shear has also entered this discussion.

Clearly the couple, or lack of couple, between any GW and HCs is a subject of immense complexity. *Warmers*, in general, seem confident the oceans will get warmer, but the skeptics would counter - maybe yes, maybe no. However, the subject of winds, and steering current, is even more complex, and no one really knows[38] what will happen to the winds.

• The "worlds most acrimonious political debate": Emanuel vs Lindzen.

Beth Daley has written[39] a rather brilliant essay that captures the essence of this odyssey. Emanuel argues why one need worry about GW. Lindzen argues why one need not worry. This site includes the basic essay, plus two videotape presentations by each scientist. The essay flips back and forth, over and over again.

Emmanuel and Lindzen are both MIT professors in Meteorology, with offices a mere one floor apart. The fact that these high horsepower scientists and long term friends disagree so vehemently may indicate that the *warmers* are getting a bit desperate.

In an Earth Day essay in the WSJ Lindzen argued "we should go back to dealing with real science and real environmental problems such as assuring clean air and water." Two weeks ago Emmanuel wrote a letter to the WSJ and called Lindzen's essay "irresponsible and misleading." He charged it was "advancing spurious hypothesis." In the end Lindzen states that "my colleague, Kerry Emanuel, received little recognition until he suggested that hurricanes might become stronger in a warmer world." The speech goes on: "He was then inundated with professional recognition."

In my view Lindzen had the better arguments. The first ten comments attached to this essay (out of 147 at time of printing) agreed, with nine for Lindzen, one neutral.

References and Notes for NOTE 9

(30) Anthes, R., et al, *Hurricanes and Global Warming — Potential Linkages and Consequences*, Bulletin, American Meteorological Society, pgs 623-628, May, 2006.

(31) Pielske Jr, R., at al, *Reply to Hurricanes and Global Warming — Potential Linkages and Consequences*, Bulletin, American Meteorological Society, pgs 628-631, May, 2006.

(32) Holland, G. & Webster, P., *Heightened Tropical Cyclone Activity in the North Atlantic: Natural Variability or Climate Trend?*, Philosophical Transactions of the Royal Society A. This rather large report was undated, but believed to be July 30, 2007.

(33) *NOAA debunks report linking hurricanes to climate change*, AXcess News, July 30, 2007. See:
http://www.axcessnews.com/index.php/articles/show/id/11810.

(34) Pielske Jr, R., *On the Holland/Webster-Landea Debate*, on Climatesci.org, August 7, 2007.

(35) Michaels, P. J. et al, *Sea Surface Temperatures and Atlantic Hurricanes*, Geophysical Research Letters, 33, 2006. See also web site CO_2 Science, May 31, 2006.

(36) Not assigned.

(37) *Loop Current*, NOAA Coastal Services Center, July 24, 2007. See:
http://www.csc.noaa.gov/crs/definitions/loop_current,html.

(38) *Gulf eddied fuel hurricanes*, USA Today Weather Focus, October 4, 2000.

(39) Daley, Beth, *A cooling trend*, The Boston Globe, via boston.com, May 18, 2010.

From Chapter 5.1

NOTE 10— Al Gore the Reincarnation of the Xhosa Prophetress Nongqawuse?

This commentary [9], by Gail Heroit, was rather outstanding. It rekindled certain memories, reinforced various convictions and reported on a con job that was unfamiliar to this writer. I will comment on: the title; the dangers of precipitous action; the treatment of our former VP as a deity or royalty; infamous prophesies by infamous prognosticators and the role/need for skepticism.

• <u>On the title</u>. This asks if Al Gore is a reincarnation of Nongqawuse? I have written that he might also be a reincarnation of Prince Albert. Canadian readers will recognize him as *the founder* of Prince Albert National Park, where he picked up his major "green credentials."

• <u>On the dangers of precipitous action</u>. Gail Heriot, is a Professor of Law at the University of San Diego School of Law, and a member of the US Civil Rights Commission. She confesses that she doesn't know much about GW, does know about "the dangers of precipitous action especially when its advocates appear to be caught up in something akin to religious fervor." She urges cautionover "the dangers of precipitous action She noted that her "instincts run toward stop, take a deep breath, and be absolutely sure that you're not about to put the world's economy in a stranglehold just to please the people who despise modernity."

• <u>On the treatment of our former VP as royalty</u>. I have written on this phenomenon also, so I was not surprised that, as a member of the Royal Family, he was given the Royal Treatment. And I liked his nickname "the Goracle." I am sorry to reflect back again on some of my writing, but this reminds me greatly of the alien creature known as "the Goron.".

• <u>The Prophetess Nongqawwuse</u>. I had not heard of this story before and her dictates. I have felt for a long time the need to bring out the ridiculous prophecies of "snake-oil peddlers", "witch doctors", and other con people.

This story is overdue. What follows are some comments on Gore and GW.

The prophet whom Gore most resembles may well be Nongqawuse She was a teenager and a member of the Xhosa tribe in South Africa. One day in early 1856, she went down to the river to draw water. When she returned, she claimed she had encountered spirits of three of her ancestors, who said her people must destroy their crops and kill their cattle. For this, the sun would rise red on February 18, 1857 and the Xhosa ancestors would sweep the British settlers from the land and bring them fresh, healthier cattle. Rather amazingly the tribal chieftain, Sarhili, agreed to do as this young girl urged. Over the next year, a killing frenzy occurred in which between 300,000 to 400,000 cattle were killed and crops destroyed. "Historians sometimes call it the Great Cattle Killing."

"On February 18, 1857 the sun rose as usual. It was not red. And the Xhosa ancestors did not show. But the Xhosa people had destroyed their livelihood. In the resulting famine, the population dropped from 105,000 to less than 27,000. Cannibalism was reported. Following Nongqawuse's advice was a calamity of staggering proportions for the Xhosa people."

Like Nongqawuse, Gore tells us the sun will soon rise red. But the models he relies on have already been proven wrong. The intense period of warming predicted by these models, over the past 10 years did not occur. We are repeatedly told it's coming, and "we must put the brakes on our use of energy—the very thing that makes the modern world possible—to avoid antagonizing the spirits of our ancestors ... I mean, to avoid climate disaster.

• Skepticism. I have also written on that subject: "Skepticism is the highest of duties for scientists, blind faith the one unpardonable sin."

References and Notes for NOTE 10
(9) Heroit, Gail, Is Al Gore the Re-incarnation of the Xhosa Prophetess Nongqawuse, guest blog on Master Resource, July 11, 2009.

NOTE 11 — On the difficulties in completing nuclear power plants
As one who spent part of his career on trying to get a Michigan nuclear project rolling—about half a dozen tiny assignments, only to see it ultimately converted to a natural gas plant— I am well aware of difficulty to get nuclear plants built.

This nuclear project triggered dozens of letters, perhaps over a hundred letters, to the local newspaper - The Midland Daily News (MDN). I may have contributed over a dozen myself, as I debated the projects merits with a Mrs. Mary Sinclair, the local anti-nuclear activist and her acolytes. Several of these letters are highlighted below.

Date	Headline	Comments
05/17/79	Papers emphasis differs	Commented on a report by the National Academy of Sciences on low level radiation. Compared U. S headlines vs MDN. Headlines
10/05/79	Observation stands	Challenged this critic's claim that "there is no net energy to be gained from nuclear power."
10/20/79	Nuclear critic mishandles data	Challenged this critic's claim that: "shows how little contribution nuclear power actually makes to the nations energy budget."
02/09/80	Nuclear critic guilty of mistakes	Responded to critic's letter that stated "Nuclear won't ease oil crunch."
04/11/80	Nuclear power right morally - - -	Argued this critic tended to carp about our society in general and nuclear power in particular. Introduced the writings of a Dr Margaret Maxey, a professor of bioethics and religious studies, to help defend the use of this energy source.
09/25/80	Get some precision in debate	Critics strived to make the case that electrical growth was over. For example in 1979 they argued that growth was a negative 0.2 percent, but that was for "total energy." For electrical energy the growth was 2.7 percent, a difference of almost three percentage points.
11/01/80	Candidate's - - leadership doubted	Challenged several letters that attempted to paint this critic as an energy leader, just before the election for state representatives. Noted six attributes that one would associate with a true energy leader. She failed in all six.

Letters to the Midland Daily News: Michigan Nuclear Project - 1979 to 1980.

With all due humility, I can claim I won this battle—Mrs. Sinclair was not elected. However, she can claim she won the war, as the two nuclear reactors were never completed. One was about 95% done and the other about 80% before the utility called it quits.

From Chapter 6.3

NOTE 12 — Hydrogen Production Chemistry

H_2 could be extracted from water via electrolysis. This would take a major amount of electricity and is a most expensive route to hydrogen. Further, any emissions caused by this new demand for electricity would need to be allocated to hydrogen.

This process is essentially 200 years old. A direct current is applied in an appropriate manner, and the following reaction occurs: $2H_2O \rightarrow 2H_2 + O_2$.

The products are released as gases.

About 4% of global hydrogen production is by this process.

In theory H_2 could also be extracted from water via electrolysis by thermo-chemical cycles. For example the Hybrid Sulfur Process[9.2], derived from a Westinghouse process, uses two reactions.

(i) A low temperature production reaction: $2H_2O + SO_2 \rightarrow H_2SO_4 + H_2$.

(ii) A high temperature regeneration reaction: $H_2SO_4 \rightarrow H_2O + SO_2 + \frac{1}{2}O_2$.

This high temperature reaction (850 - 950 ºC) would be based on energy obtained from advanced nuclear reactors. While this research is surely encouraging it must be classified as early stage.

About two days before this manuscript was sent to the publisher an excellent review[9.5] emerged. The writer noted that over 3000 thermochemical cycles have been identified, including a sulfur-iodine cycle, a variant of the Westinghouse process. This writer also noted that no thermochemical cycle "has been fully developed or demonstrated beyond the laboratory scale." However, with continued development of high temperature nuclear reactors, and the hydrogen production processes, this writer felt "commercial readiness" could be achieved shortly after 2020.

References and Notes for NOTE 12

9.2) *SRNL Hits Milestone in Nuclear Hydrogen Production*, as posted on Green Car Congress, March 10, 2009, with original date of July 6, 2007.

(9.3) Perkins, Christopher and Weimer, Alan W., *Producing Hydrogen Using Solar-Thermal Energy*, Chemical Engineering Progress, February 2009.

(9.4) Ad by Proton ENERGY SYSTEMS and SunHydro, New York Times, June 3, 2010.

(9.5) Summers, William A., *Nuclear Power: Fueling the Hydrogen Economy*, Chemical Engineering Progress, July 2010.

From Chapter 6.4

NOTE 13 — Li Ion Battery Development Status

A brief summary on the status of various battery elements follows.

(i) <u>Plastic film separator</u>. Exxon-Mobil, in conjunction with it's affiliate, Tonen Chemical have developed and is now producing[11.6] an advanced performance film for the Li-ion battery. The new films are co-extruded using specialty tailored, high heat resistant polymers. These separators offer enhanced permeability, higher meltdown temperature and melt integrity. As such the film offers significant increases in the film's thermal safety margin, and overall quality control.

(ii) <u>Silicon nano wires</u>. The key development[11.7] is the use of silicon nanowires to replace the existing carbon anode, and thereby reinvent the rechargeable lithium-ion batteries. This revised battery "produces 10 times the amount of electricity of existing lithium-ion."

The market[11.8] for Lithium-ion batteries is anticipated to grow ten fold, from 2008 to 2015, reaching $9 billion. Applications in cell phones, laptops, power tools, military and space are proving the technology for the large emerging units for hybrid vehicles and full electric automobiles.

While there has been a tidal wave of news items on the Li-ion battery, there has bee very little hard data on Li-ion battery costs, performance and lifetime. Until such data emerges this battery must be filed under a work in progress. Still, in spite of the dirth of such information, there has almost been as many news items on new demo or prototype vehicles. Examples follow.

(a) ExxonMobil Chemical reported[11.9] its new film technology will be used in Electrovaya's electric vehicles. The Maya-300 will have a range of up to 120 miles at speeds of 25-30 miles/hour.

(b) The above market report[11.8] cited about a dozen of major corporations involved in materials science and battery engineering including GE, LG Chem (see GM,below), Samsung and Sony.

(c) Reva, The Indian electric car manufacturer of the iconic *G-Wiz*, unveiled[11.10] new Li-ion battery technology that would boost the range 50% to 75 miles.

(d) Toyota announced[11.11] delivery of 500 Prius PHV's, powered by Li-ion battries, in late 2009. This new Prius was designed to use either the Li-ion battery pack, with plug in capability, or the nickel-metal hydride battery for the conventional gas-lectric system.

(e) BMW also announced[11.11] it will lease all-electric Mini-Coopers in America by March, 2009.

(f) GM announced[11.12] plans to develop and assemble Li-ion batteries in Michigan, based on cells from LG Chem. These batteries will be used in the GM Volt and other next generation vehicles. The unit for the Volt is

6 feet long and weighs 400 pounds. The Volt is designed to plug into wall outlet and be able to travel 40 miles on battery power alone.

References and Notes for NOTE 13

(11.6) ExxonMobil news release, *ExxonMobil Chemical's New Generation of Lithium-ion Battery Separator Now Produced on a Commercial Line*, Ma8, 2007.

(11.7) Stober, Dan, *Nanowire battery can hold 10 times the charge of existing lithium ion battery*, Stanford Research, December 18, 2007.

(11.8) PRLog (Press Release), *Worldwide Nanotechnology Thin Film Lithium-Ion Batey Market, Strategies and Forecasts, 2009-2015,* February 16, 2009.

(11.9) MB-BigB, *Exxon to help power new hybrid cars*, March 12, 2008.

(11.10) Murray, James, G-wiz debuts lithium-ion model, Business Green, January 6, 2009.

(11.11) environmental LEADER, Lithium-Ion Prius to arrive in Late 2009

(11.12) Krisher, Tom, *GM to put together its own batteries*, Houston Chronicle, January 13, 2009.

From Chapter 7.2

NOTE 14 Theodore Roosevelt and Progressivism.

What follows is a rather incredible odyssey of inputs on this subject.

· The first[11.1] of these was entitled: *Glenn Beck Should Revere Theodore Roosevelt*, on a Tom Woods site. The intriguing title was from a Newsmax column. However, Woods was rather dismissive of this piece - "it is written on a third grade level."

· Next[11.2] I looked at a full essay by Dr. Thomas Woods Jr. entitled: *Truth to Executive Power.* This was based on a book by Jim Powell, with main title: *Bully Boy.* Wood reports that "Powells study of TR is truly withering." Wood argues that this "critique is particularly refreshing given the cross-ideological adulation that TR has enjoyed for a full century." Wood also reports on Powell's book that: "Bill Clinton once referred to Theodore Roosevelt as his favorite Republican president. "And no wonder: TR's presidential activism, his frequent use of executive orders to effect policy - - - make him appealing to present-day Democrats and Republicans alike. "Clinton went so far as to award TR a posthumous Congressional Medal of Honor."

· Next I extracted a review[11.3] of the book: *Bully Boy: The Truth About Theodore Roosevelt's Legacy,* from Amazon.com. The first comment on this site, from Publishers Weekly, noted that Powell, a senior fellow at the Cato Institute, has made a name for himself writing provocative studies of presidents (*FDR's Folly* and *Wilson's War*). Wood in turn dismisses all inputs from Publishers Weekly.

· Then I doubled back[11.4] to the Newsmax source for the above column, by Christopher Ruddy. He acknowledges that Beck is right, on most issues, but he argues he is wrong on stating that TR "is not the root cause of President Obama's intrusive 'big-government policies'". Ruddy further argues that the reason that TR has become a controversial figure is largely due to Beck, who charges "TR was largely responsible for the 'progressive encroachments' we are seeing today."

Ruddy also acknowledges that he was, and is, an ardent TR devotee, and a member of the Theodore Roosevelt Association (TRA). He notes that Karl Rove is also a member.

· In parallel with Carl Ruddy's column, his brother, Daniel, has just had a book released[11.5] on *Theodore Roosevelt's History of the United States*. Note that Edmund Morris—who has studied Roosevelt's life for 30 years (see References 1a and 2a)—wrote the *Forward*. This book also was praised by Douglas Brinkley.

While acknowledging that Beck was correct in stating that TR was among America's first progressives, he noted that "to accurately portray TR today, we must widen our perspective and see him in the full context of his life and times." While TR embraced a progressive agenda, "the policies advocated by TR were not those of some social engineer who wanted to remake the United States based on a Saul Alinsky radical model."

· And finally[11.6] an essay entitled: *Why Mark Levin Hates Glenn Beck.* According to this columnist, Beck called TR a "precursor to what historians call 'liberal internationalism'", a foreign policy view that contends the role of the U.S. is to intervene around the globe to advance liberal objectives." He went on: "This progressive doctrine, later called 'Wilsonian' after Woodrow Wilson, was intended 'to make the world safe for democracy'." As Beck notes: there is very little difference between 'neo-conservative' foreign policy and 'liberal internationalism', and both views are progressive in origin."   Now neo-conservative talk host, Mark Levin, who "preferred to keep his audience in the dark on such distinctions" was "angry that Beck would dare shine a light on them."

In closing one might conclude it is too early to finalize an assessment of this issue.

References and Notes for NOTE 14

(11.1) Woods, Tom, "*Glenn Beck Should Revere Theodore Roosevelt*", May, 2010. See:
http://www.thomasewoods.com/blog/glenn-beck-should-revere-theodore-roosevelt/.
(11.2) Woods, Thomas E. Jr., *Truth to Executive Power*, a review of the book: *Bully Boy*, The American Conservative, November 20, 2006.
(11.3) Powell, Jim, *Bully Boy: The Truth About Theodore Roosevelt's Legacy*, Crown Forum, August 8, 2006.
(11.4) Ruddy, Christopher, *Glenn Beck Should Revere Theodore Roosevelt*, Newsmax.com, May 3, 2010.
(11.5) Ruddy, Daniel, *Theodore Roosevelt's History of the United States: His Own Words, Selected and Arranged by Daniel Ruddy*, Smithsonian, May 4, 2010.
(11.6) Hunter, Jack, *Why Mark Levin Hates Glenn Beck*, Southern Avenger, Charleston City Paper, September 25, 2009.

From Chapter 7.3

Note 15 — The Ban on DDT and the Impact on Malaria Control.

Details on this complex and controversial subject follow.

(1) <u>Applications of DDT.</u> This effort against DDT was pushed across all applications without any regard for the relative benefits involved. The fact that DDT may have been applied indiscriminately in forest and agriculture applications dominated the debate and legislation development. The early application of DDT to farm/forest spraying as compared to spraying a tiny amount of this insecticide on mosquito nets or the interior walls of each dwelling, were literally and figuratively, worlds apart. The farm/forest spraying[15], from low flying planes, used major quantities of DDT sprayed over large external areas. In contrast malaria control entails spraying a tiny amount of this insecticide on the interior walls or mosquito nets of each dwelling.

(2) <u>Driving Forces behind the Ban.</u> This effort against DDT was pushed aggressively by the Environmental Defense Fund (EDF), an NGO formed by the Audubon Society to do its dirty work for it. It could lobby for

the Audubon's objectives, without risking the Audubon's tax exempt classification. Edwards and Milloy commented on the role of the EDF as follows. "The environmental movement used DDT as a means to increase their power. Charles Wurster, chief scientist for the Environmental Defense Fund commented[7, 17] , 'If the environmentalists win on DDT, they will achieve a level of authority they have never had before'." As noted above the DDT ban was signed in 1972 on June 30th, by William Ruckelshaus, after seven months of, hearings. Apparently he never attended[18] any of these hearings, and even worse, ignored the findings of an EPA administrative law judge[7, .16] .

· DDT is not a carcinogenic hazard to man. It does not cause cancerous growths.

· DDT is not a mutagenic hazard to man. It does not induce mutations in DNA or living cells.

· DDT is not a teratogenic hazard to man. It does not cause growth abnormalities

· DDT, under the registrations involved here, does not have a deleterious effect on freshwater fish, estuarine organisms, wild birds, or other wildlife.

And still he signed the ban. Later it was reported[18] that Ruckelshaus was a member of the EDF and had even solicited donations for the EDF on his personal stationary.

(3) Malaria Economics. Some claim that other insecticides can be used. I will provide inputs from 1972 and 2002. Note that there has been no attempt to make these economic studies complete and consistent between the two time periods. They are provided, as examples, that there were serious cost problems with other pesticides in 1972 and there still are problems in the 2002 era.

· 1972 Era Costs. Less persistent, insecticides may be acceptable as a substitute for DDT in developed countries, but the poor countries of the world are faced with a huge problem, the size of which has not been appreciated. The closest alternative is Malathion, but cost is 3X that of DDT.

· Contemporary Costs. No other pesticide can be used as cost effectively as DDT. For example the Carbamates, such as Bendiocarb, are 22 times more expensive than DDT in an undissolved state and four times more expensive once applied.

(4) Unnecessary Deaths? It is necessary to raise this subject of unnecessary deaths via the ban on DDT. Could this ban be related to interests in

population control? Inputs have been obtained from a variety of sources, but only one will be noted here, as it is a charge against the EDF. It has been reported that a Charles Wurster, chief scientist for the Environmental Defense Fund commented[20], "when he was asked if people might die as a result of the DDT ban: 'Probably - - - so what? People are the causes of all the problems; we have too many of them. We need to get rid of some of them and this is as good a way as any.'" If these charges are true it brushes aside any moral dilemma, at least by this member of the EDF. It does, however. raise the possibility that the EDF may see DDT as a necessary means of population control.

References and Notes for NOTE 15

(7) Edwards, J. Gordon and Milloy, Steven, *100 things you should know about DDT,* ©1999. See:

www.junkscience.com/ddtfaq.htm

(15) The campaign against DDT and air born spraying certainly would not be hurt by the rather dramatic spraying scene, in the highly regarded 1959 movie North by Northwest where Cary Grant was *chased* by a ground hugging bi-plane that repeatedly and deliberately sprayed him before it crashed. This rather terrifying scene may well have contributed to the overall paranoia on the use of insecticides.

(16) Sweeney, E. M., *1972. EPA Hearing Examiner's recommendations and findings concerning DDT hearings,* April 25, 1972 (40 CFR 146.32, 113 pages). Summarized in Barrons (May 1, 1972).

(17) See the Seattle Times, October 5, 1969.

(18) Milloy, Steven, *Rethinking DDT,* Fox News Online, June 20, 2002. See also:

www.cato.org/cgi-bin/scripts/printtech.cgi/dailys/06-27-02.html.

(19) Tren, Richard and Bate Roger, *Malaria and the DDT Story,* The Institute of Economic Affairs, London, in association with Profile Books, 2001.

(20) Seavey, Todd, *The DDT Ban Turns 30 - Millions Dead of Malaria Because of Ban, More Deaths Likely,* American Council on Science and Health, June 2002.

From Chapter 8.1

Note 16 — Key Assumptions on Global Warming.

There are four key assumptions on the global warming issue.

(1) <u>Our planet is warming.</u>   Detection of warming is hard.

· <u>GBDs</u>. Elaborate surface temperature databases have been assembled with data from thousands of global weather stations. However, these stations lack identical instrumentation, station setting and environment. Further, far more of their measurements are based on land, far more of their measurements are in the northern hemisphere, and far more of their measurements are in developed areas, where *urban heat-island effects* (UHI) are a factor. Hence, the distribution of stations is far from uniform. Such GBDs indicate temperature is about 0.6ºC warmer today versus 100 years ago.

· <u>SDBs</u>. As an alternative approach to surface databases, a database of temperatures measured from satellites has emerged. While over 20 years old, it's coverage of our planet is 100%, and far more uniform. This database shows there has been only a tiny, if any, warming over its history.

(2) <u>This warming is caused by society</u>. The proponents of this issue assume society is guilty. But there are as many reasons to point to Nature. We live in an ocean of cyclical phenomena. The daily and annual cycles are the most obvious. Other cycles include the 2-7 year El Niño events, the ~11 year Sun-spot cycle, longer solar cycles of 80 to 1500 years, and very long term cycles of 19,000 to 100,000 years. While our understanding of these cycles is embryonic, it is rapidly improving, including understanding of their couple to climate. Hence finding society guilty is surely premature.

Proponents rely on computer models to make their case, the so-called GCMs. What is frequently ignored or left unpublicized is the quality of these models (See §3.3). Dr. William Gray, in a 1997 Houston speech commented: "when modelers move out onto the climate area, the complexity becomes too damn much[3]."

Dr. Richard Lindzen, the Sloan Professor of Meteorology at MIT, has been one of the leading critic of the GCMs. His concerns include what he considers inappropriate treatment of water vapor and cloud cover. He notes that these models only predict a significant warming when a water vapor feedback mechanism is incorporated. Yet the physics of this feedback[4] is essentially unknown. Without this feedback, increased levels of CO_2 will not lead to the dramatic warming predicted.

(3) <u>This warming will be catastrophic.</u> Well not likely.

First of all there will be benefits such as lower fuel bills, increased crop yields and increased forest growth.

Secondly, climate history over the past 1200 years, as noted in Chapter 10, of my first book, would suggest that a warming period is not all that bad, but watch out for cooling periods.

Finally the proponents, according to Dr. Stephen Schneider[5], of Stanford "have to offer up scary scenarios, make simplified, dramatic statements, and make little mention of any doubts". Hence, we see Time magazine[6] showing a picture of planet Earth, in a frying pan, on it's cover.

(4) <u>Society knows what to do to prevent this warming</u>. Well not likely.

For example Dr. James Hansen of NASA, in 1988, became the *father* of this issue, with his testimony before congress that society-caused global warming was essentially here. But in 1998 he confessed that "the *forcings* that drive long term climate change are not known with an accuracy sufficient to define future climate change[7]." Based on this confession we **should close this issue down** or at least put it on hold for 10 to 20 years.

Another example is Dr. Tom Wigley of the National Center for Atmospheric Research. He published[8] the results from his latest computer runs in 1998, again as noted in Chapter 3.1, and found the Kyoto Protocol(KP), if fully implemented by all involved nations by 2010—an event that would seem impossible to achieve—would reduce warming 0.07°C by 2050, and another 0.13°C by 2100. These amounts are so minuscule as to be unmeasurable. This means societies are being asked to spend trillions, on a course of action that we won't know is ever doing any good.

Further, assuming the *warmers* science is valid, we would have to implement not just one KP, but **about 15 successive emission cuts** to prevent a warming of say 3°C (assume 0.2°C per cut times 15 cuts). But what is debated today is just the first cut. No mention is made of any subsequent cuts required. Yet all the countries involved would have a terrible, if not impossible task, in meeting the first KP cuts alone.

References and Notes for NOTE 16

(3) Gray, W., Colorado State University, *Predicted Hurricane Activity for 1997: Is Global Warming Causing More and Bigger Hurricanes?*, Speech at the National Hurricane Association meeting, Houston, TX, April 25, 1997.

(4) Lindzen, Richard S., *The Origin and Nature of the Alleged Scientific Consensus*, Regulation, The Cato Review of Business & Government, Spring 1992.

See also: Lindzen, Richard S., *Absence of Scientific Basis*, Research & Exploration, A Scholarly Publication of the National Geographic Society, Spring, 1993.

(5) Schneider, Stephen, Discover Magazine, October, 1989.

(6) Time Magazine, April 9, 2001. The cover picture shows a black frying pan with an egg half cooked. The yoke of the egg depicts a globe and shows North America and most of South America. Cover title: Global Warming. Subtitles: Climbing Temperatures. Melting Glaciers. Rising Seas. All over the earth we're feeling the heat. Why isn't Washington?

(7) Hansen, James, *Climate Forcings in the Industrial Era*, Proceedings of the National Academy of Sciences, October 27, 1998.

(8) Wigley, T. M. L., *The Kyoto Protocol: CO_2, CH_4 and climate implications*, Geophysical Research Letters, 25, 2285-2288, July 1, 1998.

BIBLIOGRAPHY

Adamy, Janet and Laura Meckler, *Vote by Vote, a Troubled Bill Was Revived*, Wall Street Journal, March 22, 2010.

Anthrop, Donald, *Biomass potential*, Oil & Gas Journal, September 5, 2005.

—*Electric Vehicles and Carbon Emissions*, Contra Costa Times, May 22, 2010.

—*Hydrogen's Empty Environmental Promise*, Cato Institute, Paper No. 90, December 7, 2004.

—*The U.S. Carbon Emissions, and the Kyoto Protocol*, a point and counterpoint presentation, on climate change with Anthrop taking the counterpoint, Pacifica, Spring 2006.

Begley Sharon, et al, *The Truth About Denial*, Newsweek, August 13th, 2007.

Bentley, C. R., *Rapid sea-level rise from a West Antarctic ice-sheet collapse: a short term perspective*, J. of Glaciology, Vol. 44, no. 146, pp 157-163, 1998, as posted on CSA Illumina.

Berger, A., *Introduction to the Milankovitch Theory of Climate", Review of Geophysics, 26, November 1988*.

Blyth, Myrna, Spin Sisters - *How the Women of the Media Sell Unhappiness and Liberalism to the Women of America*, St. Martins Press, New York, NY, March 2004.

Bradley, Robert L., *Corporate Social Responsibility and Energy: Lessons from Enron*, Institute for Study of Economics and the Environment, Lindenwood University, April 2008.

Bryant, Edward, *Climate Process & Change*, Cambridge University Press, 1997.

Bryce, Robert, *Gusher of Lies - The Dangerous Delusions of "Energy Independence"*, PublicAffairs™, 2008.

Boswell, Randy, *Noah's Flood: the Canada connection. Scientist Links Lake Agassiz To Noah's Flood*, Winnipeg Free Press, May 9, 2004.

Buchmann, Isidor, *Is lithium-ion the ideal battery*, Created: April 2003, last edited: Nov. 2006.

Campbell, Ellen, *Cambridge Who's Who names James O'brien Professional of the Year in Meteorology, April 18, 2007.*

Campoy, Ana, *Betting on a Biofuel*, Wall Street Journal, June 30, 2008.

Carson, Rachel, Silent spring, Houghton Miffin Company (2002), Boston.

Center for Strategic and International Studies, Georgetown University, *Where We Agree - Report of the National Coal Policy Project*, Westview Press, Boulder, CO, 1978.

Chernova, Yuliya, *Shedding light on Solar*, Wall Street Journal, June 30, 2008.

Church, J. A., *Estimate of the regional distribution of sea level rise over the 1950-2000 period*, J. of Climate, 17, 2004, p, 2609-2625.

Climate Model Inadequacies (Clouds) - Summary, Last updated January 25, 2006

Consequences, Bulletin, American Meteorological Society, pgs 628-631, May, 2006.

Colburn, Theo, et al, *Our Stolen Future: Our We Threatening Our Fertility, Intelligence and Survival. A Scientific Detective Story,* Dutton Publishing, 1996.

Cotterly, W., *Hurricanes & Tropical Storms: Impact on Maine and Androscoggin County*, 1996

Daley, Beth, *A cooling trend*, The Boston Globe, via boston.com, May 18, 2010

Daly, Brenda, Nadear Elshami, *Pelosi, Reid Call on Bush to Support mandatory Limits on Greenhouse Gases*, Speaker Nancy Pelosi press release, September 28, 2007.

Davidson, Paul, *New battery packs powerful punch*, USA Today, July 4, 2007.

Deutch, John and E. Moniz, *A Future for Fossil Fuel*, Wall Street J., March 15, 2007.

Dunetz, Jeff, *Senate Cap and Trade Skipping Committee Process - Will Be Drafted Behind Closed Doors*, RedSate.com: Conservative Blog and News, April 15, 2010.

Economides, Michael, *Propaganda as Journalism*, Human Events, August 20, 2007.

Editorial, *The first step: House pasage of energy bill marks new U. S. willingness to fight climate change*, Houston Chronicle, July 1, 2009.

Editorial: Is *the House Swamp Drained Yet?* The New York Times, April 18, 2009.

Editorial: Environmental News Network staff, *West Antarctic Ice Sheet not in jeopardy*, CNN

Interactive, December 1, 1998. See Internet, www.cnn.com/TECH/science/9812/01/

Editorial, *Green Power*, Fortune Small Business, April 2009.

Editorial: *Backward California*, Oil & Gas Journal, February 23, 2009.

Editorial: *Cap and Trade Legislation: Will California's AB32 Go National?*, February9, 2009.

Edwards, J. Gordon, *The lies of Rachel Carson*, eco-logic on-line, November 1, 2002.

Editorial: Is the House Swamp Drained Yet?, The New York Times, April 18, 2009.

Eilperin, Juliet, *Hackers steal electronic data from top climate research center*, The Washington Post, November 21, 2009.

Espo, David, *Pelosi Says She Would Drain GOP 'Swamp'*, The Washington Post, October 6,

Evers, Marco et al, Climate *Catastrophe - A Superstorm for Global Warming Research*, Spiegal Online International, April 1, 2010.

Fara, P., *Learning from the Past*. See the Global Warming Debate - The Report of the European Science and Env. Forum, edited by J. Emsley, Bourne Press Ltd, Bournemouth, Dorset, 1996.

Fairbanks, R. G., *A 17,000 year glacio-elastic sea level record: influence of glacier melting rates on the Younger-Dryas event and deep ocean circulation*, Nature **342**, December 7, 1989.

Farah, Joseph, *Al Gore and the new journalism*, WorldNetDaily, February 28, 2001.

Friedman, Brad, *RNLA Issue Fact-Free Letter Claiming Desperate Franken is Stealing U.S. Senate Seat*, February 15, 2009. See http://bradblog. com/?p=6918.

Gallego, J., C. Cabellero, *Green Energy "Has Cost Many Jobs,"* El Mundo, May 23, 2009.

Glassman, James K., *Hurricanes and Global Warming: Interview with Dr. James J. O'Brien,* Capitalism magazine, September 13, 2005.

Gore Jr., Albert, *To Skeptics on Global Warming*, New York Times, April 22 1990.

—*Earth in the Balance*, Houghton Mifflin Company, New York, NY, 1992.

—*An Inconvenient Truth*, Rodale, Inc., Emmaus, PA, May 26, 2006.

Fowler, Tom, *A role for coal, new drilling*, Houston Chronicle, February 10, 2009.

Graumann, Axel, et al, *Hurricane Katrina - A Climatological Perspective - Preliminary Report*, Technical Report 2005-01, NOAA'S National Climatic Data Center, October, 2005.

Gray, W., et al, *Early April Forecast of Atlantic Basin Seasonal Hurricane Activity for 1997*, Dept of Atmospheric Science, Colorado State University, April 4, 1997.

Grunwald, Michael, *The End of California? Dream On!*, Time, November 3, 2009.

Hansen, James A., et al, *Dangerous Human-Made Interference with Climate*, Atmos. Chem. Phys. Discussions., December 5, 2006, March 29, 2007, and May 7, 2007.

Hansen, James et al, *Global Climate Changes as Forecast by Goddard Institute for Space Studies Three Dimensional Model*, See Appendix B: Radiative Forcings, Journal of Geophysical Research, August, 1988.

Hansen, James A., *Dangerous Human-Made Interference with Climate*, Testimony before the U. S. House of Representatives, April 26, 2007.

Hansen, James, *Climate Forcings in the Industrial Era*, Proceedings of the National Academy of Sciences, October 27, 1998.

Hensen, Bob, *rough Sea regional variations add a wild card to future sea-level rise*, UCAR Magazine, Fall, 2009.

Editorial, Environmental News Network staff, *West Antarctic Ice Sheet not in jeopardy*, See: Internet, www.cnn.com/TECH/science/9812/0/ CNN Interactive, December 1, 1998.

Hayes, W., 1956. Journal of the American Medical Association (JAMA) 162: 890 - 897.

Hazeltine, W. E., 1972. *Why pelican eggshells are thin*, Nature, 239, pp 410 - 412.

Heidorn, Kieth, *The 1900 Galveston Hurricane*, The Weather Doctor, September 1, 2000.

Heinlein, Jr., Robert, A., *The Green Hills of Earth*, a short story, with the poem of the same title, Oct 20,1999. See: http://www.cs.rice. edu/~ssiyer/minstrels/poems/241.html.

Helvarg, David, *Fiddling While Antarctica Burns*, New York Times, March 7, 1999.

Henderson, Laura, C. Tucker, *Former president channels Prof. Gabriel Cazada in delivering veiled rebuke of Obama's Spanish-inspired green jobs plan*, posted on The Institute for Energy Research, May 27, 2009.

Heriot, Gail, *Is Al Gore the Re-incarnation of the Xhosa Prophetess Nongqawuse*, guest blog on Master Resource, July 11, 2009.

Herszenhorn, David, *Senate Closes Down Climate Change Bill*, NY Times, June 7, 2008.

Holland, G. & Webster, P., *Heightened Tropical Cyclone Activity in the North Atlantic: Natural Variability or Climate Trend?* Philosophical Transactions of the Royal Society A., 1998.

Horner, Christopher C., *The Politically Incorrect Guide™ to Global Warrming and Environmentalism*, Regnery Publishing Company, Washington, DC, 2007.

Hoyt, Douglas V., et al, *The Role of the Sun in Climate Change*, Oxford University Press, 1992.

Huber, Kathy, *Your nose will know*, Houston Chronicle, July 8, 2010.

Huber, Peter, *Hard Green - Saving the Environment from the Environmentalists*, Basic Books, New York, NY, 1999.

Hunter, Jack, *Why Mark Levin Hates Glenn Beck*, Southern Avenger, Charleston City Paper, September 25, 2009.

Idso, Sherwood and Idso, Craig, *Carbon Dioxide and Global Change: Separating Scientific Fact from Personal Opinion*, June 6, 2007.

Idso, Sherwood B., *CO2 induced global warming: a sceptic's view of potential climate change, Climate Research*, 10: 69-82, 1998.

—*Letters*, Carbon Dioxide and Climate, Science, 210, 7-8, 1980.

—*The Climatological Significance of a Doubling of Earth's Atmospheric Carbon Dioxide Concentration*, Science 207: 1462-63, 1980.

—*Weather Extremes (Precipitation - Model Inadequacies) - Summary*, Last updated May 4, 2008. See www.co2science.org.

Idso, Sherwood B., *CO2 Warming is Good for the Planet*, New York Times, May 7 1990.

—*Letters: Carbon Dioxide and Climate*, Science, 210, 7-8, 1980.

IPCC, Climate *Change 2007 - The Science of Climate Change*, Cambridge U. Press, (2008).

—Climate *Change 1995 - The Science of Climate Change*, Cambridge U. Press, (1996).

—*A Report of Working Group I of the Intergovernmental Panel on Climate Change, Summary for Policymakers*, 2007, Figure SPM.2..

Jastrow, R., W. Nierenberg, and F. Seitz, *Scientific Perspectives on the Greenhouse Problem*, The Marshall Press, Jameson Books Inc., Ottawa, IL, 1990.

Karl, T.R., et al, *Asymmetric Trends of Daily Maximum and Minimum Temperature*, Bulletin of the American Meteorological Society, **74**, 1993

Keeling, C. D. et al, *Atmospheric Retention of CO_2*, Nature, **375**, 6-22-95

Kerr, Richard, *Model gets it right - without fudge factors*, Science, May 16, 1997.

—*Dark Clouds Promise Brighter GCM Future*, Science, **267**, 1-27-95

—*Climate Modeling's Fudge Factor*, Science, **265**, 9-9-94

Khandekar, Madhav L., *Questioning the Global Warming Science: An annotated bibliography of recent peer-reviewed papers, Section 2 - Impact of solar variability on the earth's climate*, Friends of Science, January 2007.

Leake, J. and C. Hasting, *World misled over Himalayan meltdown*, The Sunday Times, January 17, 2010, as posted on Timesonline.

Johnson, Keith and Steven Chu, *'Coal is My Worst Nightmare'*, The Wall Street Journal, as reported in http://blogs.wsj.environmentalcapital - - -, December 11, 2008.

Johnston, M. and T. Holloway, *A global comparison of national biodiesel production potentials*, Environmental Science & Technology, 41 (23), 2007.

Klein, Ezra, *Nancy Pelosi on Health-Care Reform*, The Washington Post, July 22, 2009.

Kohl, Kieth, *Natural Gas Production*, Energy and Capital, September 15, 2008.

Kolata, Gina, *Study Discounts DDT Role in Breast Cancer*, New York Times, October 30, 1997.

Krauthammer, C., *Gore's campaign lived, died by courts*, Houston Chronicle, Dec. 15 2000.

—*The Fierce Urgency of Pork*, The Washington Post, February 6, 2009.

Krisher, Tom, *GM to put together its own batteries*, Houston Chronicle, January 13, 2009.

Lammers, Dick, *Can ethanol be the fuel of the future?*, Houston Chronicle, February 15, 2009.

Lanier, Elyse, *It would help if Houstonians stopped selling our city short*, Houston Chronicle, April 8, 2001.

Larson, Erik, *Isaac's Storm*, Crown Publishers, New York, 1999.

Letourneau, J., *Where is the North American Natural Gas Market Headed*, Seeking Alpha, June 15, 2007.

Levin, Marc, *Democratic Spokesman Says "Houston is Filthy"*, Houston Review, September-October 2000. See: www.houstonreview.com/articles/FilthyHouston.html.

Lewis Jr., Marlo, *Al Gore's: An inconvenient Truth*, See Competitive Enterprise Institute publication: On Point, September 28, 2006

Lifsher, Marc, *Key points still debated regarding California renewable energy goals*, Los Angeles Times, August 29, 2009

Lindzen, Richard, Global Warming Facts – *Some Relevant Figures for Current Behavior of Global Mean Surface Temperature*, See www.ecoworld.com, October 15, 2006.

—*Response*, BAMS, 71: 1465-1467, 1990.

—*Absence of a Scientific Basis*, National Geographic Research & Exploration, 9(2), 1993

—*Some Coolness Concerning Global Warming,*. BAMS, 71: 288-99, 1990.

—*Climate of Fear*, The Wall Street Journal, April 12, 2006.

—*Errors Hurt Global Warming Theories*, NY Times, 11-30-90.

—*The Origin and Nature of the Alleged Scientific Consensus*, Regulation, The Cato Review of Business & Government, Spring 1992.

—*Climate Alarm: What We Are Up Against, and What to Do*, Heartland Institute Conference, March 8, 2009.

—*Some uncertainties with respect to water vapors role in climate*, 1990.

—*Global Warming Facts – Some Relevant Figures for Current Behavior of Global Mean Surface Temperature*, See www.ecoworld.com, October 15, 2006.

Llewellyn, Richard, *How Green Was My Valley*, Macmillan Publishers, 1940.

Lockwood, M., et al, *A Doubling of the Sun's Coronal Magnetic Field during the Last !00 Years, Nature, June 3, 1999*.

Lomborg, Bjørn, *Global warming will save millions of lives*, The Telegraph, March 12, 2009.

—*the skeptical environmentalist - Measuring the Real State of the World*, Cambridge University Press, 2001.

Malkin, Michelle and Michael Fumento, *Rachel's Folly - The End of Chlorine*, Competitive Enterprise Institute, 1996.

McCarthy, Michael, *The Skeptical Environmentalist by Bjørn Lomborg, A cool head in the hot air*, THE INDEPENDENT, August 31, 2001.

McGranahan, G. and D. Satterthwaite, *Is the 'Green Agenda' appropriate in poor cities?*, as posted at webmit.edu, December 5, 2000 *Environmental Health or Ecological Sustainability: Reconciling the brown and green agendas in urban development.*

McGurn, William, *Pelosi's Indefensible Bill*, The Wall Street Journal, February, 10, 2009.

McQueen, Ian, *Sing a Song of Turbines*, Telegraph-Journal, Canada East, June 20, 2008.

Meyers, Jim, *Inhofe: Cap-and-Trade Has No Chance in Senate*, Newsmax.com, April 25, 2010.

Michaels, P. J. et al, *Sea Surface Temperatures and Atlantic Hurricanes*, Geophysical Research Letters, 33, 2006. See also web site CO_2 Science, May 31, 2006.

Mitchell, J.F.B., et al, *On Surface Temperature, Greenhouse Gases, and Aerosols: Models and Observations*, Journal of Climate, **8**, 10-95

Milloy, Steven, *Rethinking DDT*, Fox News Online, June 20, 2002.

Milne, Lorus and Margery, *There's Poison All Around Us*, New York Times Book Reviews, September 23, 1962.

MIT News Release, The Future of Coal,, March 14, 2007. This news release is for the report: *Future of Coal - Option for a Carbon Constrained World.*

Morris, Edmund, *The Rise of T. Roosevelt*, Coward, McCann & Geoghagen, New York, 1979.

—*Theodore Rex*, Random House Inc., New York, 2001.

Mouawad, Jad, *ExxonMobil - Green is for Sissies*, New York Times, November 16, 2008.

Mouawad, Jan and Diana R. Henriques, *Why is Oil So High? Pick a View*, The New York Times, June 21, 2008.

Murray, James, *G-wiz debuts lithium-ion model*, Business Green, January 6, 2009.

Murty Tad, *Katrina and History*, Montreal Gazette, September 1, 2005.

National Coal Council, *Coal: Americas Energy Future*, Industrial Environment, May 1, 2006.

NCAR: *When Models and Satellites Mislead*, March 13, 1997. This report covers two papers published in Nature of the same date. The key quote by Kenneth Trenberth is in the second paper: *The use and abuse of climate models.*

NOAA debunks report linking hurricanes to climate change, AXcess News, July 30, 2007. See:

NOAA, See: Loop Current, Coastal Services Center, July 24, 2007,

Pelosi, Nancy, *We Must Pass The Health Care Bill* So That We Can See What's In It.

O'brien, James, *Atlantic Hurricanes: The True Story, Washington Roundtable on Science & Public Policy, George C. Marshall Institute, October 12, 2005.*

O'Reilly, Bill, *Polls Reveal Viewer Opinions on Media, an interview with Bernard Goldberg*, The O'Reilly Factor, as reported on LexisNexis News home Page, November 2, 2009.

Pielske Jr, R., at al, *Reply to Hurricanes and Global Warming — Potential Linkages and Consequences*, Bulletin, American Meteorological Society, pgs 623-628, May, 2006.

Perkins, Christopher and Alan W. Weimer, *Producing Hydrogen Using Solar-Thermal Energy*, Chemical Engineering Progress, February 2009.

Philpott, Tom, *New Energy Secretary Chu: Big proponent of cellulosic ethanol*, BioDieselNow, December 17, 2008.

Pielske Jr, R., *On the Holland/Webster-Landsea Debate*, Climatesci.org, August 7, 2007.

Poor, Jeff, *Iwo Jima Veterans Blast Time's 'Special Environmental Issues' Cover*, Business & Media Institute, April 17, 2008.

Powell, Jim, *Bully Boy: The Truth About Theodore Roosevelt's Legacy*, Crown Forum, August 8, 2006.

Priore, Suzanne, *AEP Dedicates First Use of Stationary Sodium Sulfur Battery*, American Electric Power, September 23, 2002.

Randall, Tom, *Uncle Sam: Sell that land,* Environmental & Climate News, March 2000.

Rapley, C., Interview: *Scientist rebuts global warming critics,* British Antarctic Service, Houston Chronicle, February 7,.2006.

Rapier, Robert, *Mythbusters: Ethanol and Foreign Oil Displacement,* The Oil Drum, as reported on The Intelligence Daily, August 8, 2008.

Report: Hurricane Alicia, 1983, USA Today, August 30, 1999.

Report: The Great Flood of 2001, Texas - Houston Chronicle Magazine, July 15, 2001.

Rhodes, Kathleen, *Liberal Bloggers Pounce on Voting Fraud Watchdog Group,* CNSNEWS, March 31, 2005.

Roosevelt, Theodore, *Hunting Trips of a Ranchman,* G. P. Putnam and sons, New York, 1885.

Rubin, Paul H., *Environmentalism as Religion,* Wall Street Journal, April 22, 2010.

Ruckelshaus, William, *A New Shade of Green,* Wall Street Journal, April 17, 2010.

Ruddy, Christopher Ruddy, *Glenn Beck Should Revere Theodore Roosevelt,* Newsmax.com, May 3, 2010.

Ruddy, Daniel, *Theodore Roosevelt's History of the United States: His Own Words, Selected and Arranged by Daniel Ruddy,* Smithsonian, May 4, 2010.

Schneider, Stephen H., et al, *Letters: Carbon Dioxide and Climate,* Science, 210, 6-7, 1980.

Schneider, Stephen, Discover Magazine, October, 1989.

Schwartz, S., et al, *Uncertainties in Climate Change Caused by Aerosols,* Science, **272**, 5-24-96.

Seavey, Todd, *The DDT Ban Turns 30 - Millions Dead of Malaria Because of Ban, More*

Deaths Likely, American Council on Science and Health, June 2002.

See: Houston Chronicle Magazine *Report: The Great Flood of 2001, Texas,* July 15, 2001.

See: *Katherine Harris,* http://en.wikipedia.org/wiki/Katherine_Harris.

See: *Author recounts 'the deadliest hurricane in history'*, CNN book news, August 25, 1999.

See: Statement made by Richard Lindzen, Transcripts: CNN Larry King Live, *Could Global Warming Kill us?*, January 31, 2007.

See: NASA, Marshall Space Flight Center on Solar Physics, *Why We Study the Sun Space Weather*.

See Newsmakers Albert Gore Jr., ABCNews, 2000, quoting Vanity Fair magazine, March 1988.

See: *A Global Warming Snow Job?*, *at* www.worldclimatereport.com, May 27, 2005.

See: the Roger Pielke Sr. Web site at: http://climatesci.colorado.edu. Select January 2007 entry for *A Personal Call for More Modesty, Integrity and Balance* - by Hendrik Tennekes.

See: http://wwwc3headlines.com/global-warming-economicsprofiteeringreparations/. The input of interest is the 8th one on this C3 site, March 12, 2010.

See: Ironic Surrealism blogivists.com, *What was in the Waxman-Markey 'Manager's Amendent'?*, June, 2009.

See:, *Let's Laugh at California - They Deserve It*, See: A Conservative News Forum, March 27 2001. www.freerepublic.com/forum/a3ac14c4b2f25.htm.

See: http://www.cbc.ca/green/, for a 2010 CBC promotion in Canada.

See: http://www.nypa.gov/press/2009/090109a.htm. *MTA LI Bus and NYPA Install First Sodium Sulfur Battery Energy Storage System in State*, January 9, 2009.

See: http://www.thomasewoods.com/blog/glenn-beck-should-revere-theodore-roosevelt/. Woods, Tom, *"Glenn Beck Should Revere Theodore Roosevelt"*, May, 2010.

See: News release, *ExxonMobil Chemical's New Generation of Lithium-ion Battery Separator Now Produced on a Commercial Line*, May, 2007.

See: www.haas.berkeley.edu/news/california_electricity_crisis.html. *Manifesto on the California Electricity Crisis*, Institute of Management, Innovation and Organization, U. of California, 1-26-01.

See: www.junkscience.com/ddtfaq.htm. Edwards, J. Gordon and S. Milloy, *100 things you should know about DDT.*

Senior, Antonio, *Blunt warning about greens under the bed*, Times OnLine, July 24, 2009.

Shannon, Meg, *How Green is Gwyneth Paltrow*, as reported on Fox News, July 23, 2009.

Sidell, M. Et al, *Retraction: Constraints on future sea-level rise from past sea-level changes*, Nature Geoscience, July 26, 2009.

Simmons, Matthew R., *How Mature are the World's Super Giant and Giant Oil & Gas Fields and are they Still Important*, presentation, Pioneer Oil Producers Society, Houston, March 16, 2009.

—*The Oil and Gas System is Sick*, presentation at The Commercial Club of Boston, February 11, 2009.

Simon, Stephanie, *Green vs Growth: The Battle Rages On*, Wall Street Journal, April 17-18, 2010. This article also includes several partial clippings from their archives, including: *Shades of Green Eight of 10 Americans are Environmentalists, At Least So They Say*, August 2, 1991.

Simulating the Past: *A Test of State-of-the Art Climate Models*. A review of the technical paper by Lau, K. M., et al, *A multimodel study of the twentieth century simulations of Sahel drought from the 1970s to the 1990s*, Journal of Geophysical Research, 10. 2006.

Sweeney, E. M., *1972. EPA Hearing Examiner's recommendations and findings concerning DDT hearings*, April 25, 1972 (40 CFR 146.32, 113 pages). Summarized in Barron's (May 1, 1972).

Snyder, Mike, *A better Bayou*, Houston Chronicle, September 22, 2002.

SRNL Hits Milestone in Nuclear Hydrogen Production, as posted on Green Car Congress, March 10, 2009, with original date of July 6, 2007.

Stanton, S., *Special Report: How Californians got burned - - The state electrical system is in a shambles, and the worst may be ahead. How did things get to this point?*, The Sacramento Bee News, May 6, 2001.

Steffy, Loren, *Wind Whispers of Enron*, Houston Chronicle, June 2, 2008.

Steinhubl, Andrew, et al, *Unconventional resources to keep pivotal supply role*, Oil & Gas Journal, January 26, 2009.

Stober, Dan, *Nanowire battery can hold 10 times the charge of existing lithium ion battery*, Stanford Research, December 18, 2007.

Stott, Philip, *Cold Comfort for 'Global Warming'*, New York Times, March 25, 2000.

Streifeld, David, *Uprising Against the Ethanol Mandate*, New York Times, July 23, 2008.

Summers, William A., *Nuclear Power: Fueling the Hydrogen Economy*, Chemical Engineering Progress, July 2010.

Svensmark, H., et al, *Variation of cosmic ray flux and global cloud coverage ... a missing link in solar-climate relationships*, J. of Atmospheric and Solar-Terrestrial Physics, 59, 1225-1232, 1997.

Svensmark, H. and N. Calder, *The Chilling Stars*, Totem Books, March 19, 2003.

Sweeney, E. M., *1972. EPA Hearing Examiner's recommendations and findings concerning DDT hearings*, April 25, 1972 (40 CFR 146.32, 113 pages). Summarized in Barrons (May 1, 1972).

Telugu, *Tad Murty Wins Indo Canada Award*, June 11, 2005. See: http://www.tlca.com.

Terzian, P., *Bland Ambition*, The American Spectator, August 1999. This essay is based on the book - *Gore: A Political Life* by Bob Zelnick, a former ABC News correspondent. Terzian credits Zelnick's reporting as enabling him to see Gore with some clarity and to help him to penetrate the mystery of this politician

The Hurricane of '38, American Experience, © 1999 - 2002.

Titus, James G., (2004), *Maps that Depict the Business-As-Usual Response to Sea Level Rise in the Decentralized United States of America*, OECD Global Forum on Sustainable Development and Climate Change, ENV/EPOC/GF/SD/RD(2004)9/Final, OECD, Paris.

Titus, J. G. and V. K. Narayanan, *The Prob. of Sea Level Rise*, EPA 230-R95.008, October 1995.

The Hydrogen and Fuel Cell Investor, © 1999 - 2010.

Trabant, D. C., et al, *Hubbard Glacier, Alaska: Growing Faster and Advancing in Spite of Global Climate Change - - - ,* USGS, January, 2003.

Tren, Richard and Roger Bate, *Malaria and the DDT Story*, The Institute of Economic Affairs, London, in association with Profile Books, 2001.

Tyner, Sr., Gene, *Net Energy from Wind Power*, Minnesotans For Sustainability, Jan. 2002.

USA Today Report, *Hurricane Alicia, 1983*, August 30, 1999.

Vastabedian, Ralph and P. Pae, *A Barrier That Could Have Been*, LA Times, Sept. 9, 2005.

Verma, R. P., *Butanol - A possible Alternative Energy Source*, Int. Symposium on Biofuels, September 25, 2007. See http://petrofed.winwinhosting. net/uploa/4_Verma.pdf

Verneer, M. and S. Rahmstorf, *Global sea level linked to global temperature*, Proceedings of the National Academy of Sciences Early Edition, December 4, 2009.

Wadler, Joyce, *Green Guilt. Even the most committed say they commit environmental sins*, New York Times, September 30, 2010.

Wald, M. L., *Alternate Fuels: All Gallons Are Not Equal*, New York Times, May 28, 2006.

Walsh, Bryan, *'WE' Climate Campaign: Glossy, But Will It Work*, Time, September 1, 2008.

Waters, Clay, *NY Times Tackle Damming GW Emails*, Newsbusters, November 19, 2009.

Watts, Anthony, *Svensmark: "global warming stopped and a cooling is beginning" – "enjoy global warming while it lasts."* See: Watts Up With That? web site, October 9, 2009.

Weart, Spencer, *The Discovery of Global Warming*, Harvard University Press, Sept. 30, 2004.

Wegman, Edward, et al, AD HOC COMMITTEE REPORT ON THE 'HOCKEY STICK' GLOBAL CLIMATE RECONSTRUCTION, 2006. A search of the Internet will provide this and many other documents on this controversial subject.

Westbrook, Gerald T., Papers at Offshore Technology Conference, Houston, Texas

—*Incredible Story of the Worlds Oceans: Will GW Have an Impact*, #8689, May 4, 1998.

—*Global Warming and the World's Oceans - Update*, #10773, May 3, 1999.

—*Global Warming and the World's Oceans - The Millennium Outlook*, #12115, May 5, 2000.

—*The Global Warming Issue: Current Status*, #14284, May 6, 2002.

—*More on Propaganda as Journalism, including the "P, P & P Test"*, Posted initially on The ecologicPowerhouse web site, September 7, 2007.

—*Global warming Models: Are they Adequate for Policy Development?* IAEE Newsletter, Summer, 1997.

—*Global Warming: Witnesses for the Defense of the Skeptical Perspective - Physicists*, IAEE Energy Forum, 3rd quarter, 2008.

—*The Skeptics on the Global Warming Issue: The Distinguished Veterans*, IAEE Newsletter, 4th quarter, 2005.

—*'Acid Rains' on Liberal Propaganda*, iUniverse Inc., New York, 2004

—*Global Warming: Who to Believe?* AICHE, April, 2007.

—*Global Warming: Witnesses for the Defense of the Skeptical Perspective*, Energy Tribune, May 29, 2007. See www.energytribune.com/articles.cfm?aid=500.

—*Global Warming: Witnesses for the Defense of the Skeptical Perspective - Physicists*,

IAEE Energy Forum, 3rd quarter, 2008.

—*The Skeptics on the Global Warming Issue: The Distinguished Veterans*, IAEE Newsletter, 4th quarter, 2005.

—*Warming debate needed*, letter, Oil & gas Journal, November 3, 2008.

Wigley, T. M. L., *The Kyoto Protocol: CO_2, CH_4 and climate implications*, Geophysical Research Letters, 25, 2285-2288, July 1, 1998.

Will, George F., *A Quixotic Pursuit: Green Energy Jobs*, The Washington Post, June 25, 2009.

—*Opposition to ANWR drilling? It's collectivism in drag*, Houston Chronicle, December 16, 2005.

—*The President is CPR for the GOP*, Newsweek, September 21, 2009.

Woods, Thomas E. Jr., *Truth to Executive Power*, The American Conservative, Nov. 20, 2006.

INDEX